CAMBRIDGE MONOGRAPHS
ON MATHEMATICAL PHYSICS

General Editors: D. W. Sciama, S. Weinberg, P. V. Landshoff

MODELS OF
HIGH ENERGY PROCESSES

T0297273

MODELS OF
HIGH ENERGY PROCESSES

J. C. POLKINGHORNE

Department of Applied Mathematics and Theoretical Physics
University of Cambridge

CAMBRIDGE UNIVERSITY PRESS

CAMBRIDGE

LONDON   NEW YORK   NEW ROCHELLE
MELBOURNE   SYDNEY

CAMBRIDGE UNIVERSITY PRESS
Cambridge, New York, Melbourne, Madrid, Cape Town, Singapore,
São Paulo, Delhi, Dubai, Tokyo

Cambridge University Press
The Edinburgh Building, Cambridge CB2 8RU, UK

Published in the United States of America by Cambridge University Press, New York

www.cambridge.org
Information on this title: www.cambridge.org/9780521133821

First published 1980
This digitally printed version 2010

A catalogue record for this publication is available from the British Library

Library of Congress Cataloguing in Publication data
Polkinghorne, J C 1930
Models of high energy processes.
(Cambridge monographs on mathematical physics)
Bibliography: p.
Includes index.
1. Particles (Nuclear physics) – Mathematical
models. 2. Perturbation (Quantum mechanics)
3. Feynman integrals. I. Title.
QC793.2.P64   530.1'2   79–296

ISBN 978-0-521-22369-0 Hardback
ISBN 978-0-521-13382-1 Paperback

To
RUTH

Contents

Preface

Theoretical physics makes extensive use of models to test and develop intuition. In non-relativistic quantum mechanics the principal source of insight is provided by the study of suitably chosen potentials. However, such an approach can be of little value in relativistic quantum mechanics. Instead the Feynman integrals of perturbation theory have provided a rich testing ground for assessing dynamical conjectures. The method is unashamedly heuristic but it commands respect because Feynman perturbation theory gives a formal solution of the requirements of analyticity and unitarity. These principles are believed to provide the essential kinematic setting for relativistic quantum mechanics. It is true that recent ideas of confinement, and of the role of non-perturbative classical solutions of field theory, have suggested important aspects of relativistic quantum mechanics that are not to be seen in Feynman integrals. Nevertheless the method retains its power to act as a guide to the answer of many dynamical questions. In particular it remains an indispensable tool to investigate the fundamental interactions of quarks and gluons, a role which has been given an enhanced respectability by the elegant notion of asymptotic freedom for non-Abelian gauge theories.

While perturbation theory continues to be an important model, eliciting its guidance is sometimes a formidable analytic task. An important advance was made when Academician Gribov introduced hybrid models, based on Sudakov parameter methods. Not only are these models in many cases easier to calculate but also their largely non-perturbative character makes their conclusions stand on firmer ground. The technique pioneered by Gribov has proved a fruitful source of model making for many physical regimes. One of its most important uses has been to provide a covariant and non-perturbative formulation of the parton model. This model describes the substructure within hadrons which appears to be manifest in deep inelastic scattering reactions of all kinds. Such processes, characterised by high energy and large momentum transfer, probe the constituents out of which hadrons are made. They provide much of the detailed evidence for the quark structure of matter.

This monograph seeks to provide an introduction to these types of model making. Its aim is to explain the basic ideas in a form accessible to graduate students and other readers who have acquired a first knowledge of quantum field theory and basic particle physics, including the elements of Regge theory. I believe that it describes all major calculational techniques together with sufficient physical applications to illustrate their utility. No attempt has been made to be encyclopaedic, for an exhaustive treatment of every application would have created a volume too large for the simple pedagogic purpose intended. For example the parton model is discussed in a way which exhibits its physical structure but which avoids commitment to details which are still a matter of unresolved phenomenological debate. Similarly, on the theoretical side I have been content to illustrate the connection of the ideas presented with Reggeon field theory and with K. Wilson's operator product expansion, without developing either subject in detail since each is really an autonomous discipline in the regime it describes.

I am very grateful to Dr P. V. Landshoff and Mr W. J. Stirling for reading an early draft and making valuable comments, to Dr I. G. Halliday for useful suggestions, to Miss Sandra Evans for deciphering my handwriting and typing the manuscript and to Mr C. Chalk for drawing the many figures. I would also like to thank the staff of the Cambridge University Press for their help and care in the preparation of this book.

J. C. POLKINGHORNE

Summary of analytical techniques

In this monograph we describe a number of mathematical techniques. They are employed in appropriate physical contexts but often they are capable of much wider application than can be illustrated in a book of this size. The aim of this summary is to give an indication of these techniques, and the sections in which they are developed, in the hope that this will prove useful to a reader in search of a line of attack on a problem.

The basic method for evaluating Feynman integrals is symmetric integration (section 1.2). If numerator factors are present the use of auxiliary momenta as dummy variables is often helpful (section 1.3). A technique for handling logarithmic factors is known (section 3.3, equation (3.3.28)). Sometimes it is convenient to rewrite the loop momentum integrals as integrals over invariants (section 3.7). Ways of handling θ-functions and δ-function constraints are also available (section 4.4).

The asymptotic behaviour of integrals can sometimes be determined by direct integration by means of formulae like (2.1.7) of section 2.1. This section describes the important notions of natural behaviour, end point contributions and pinch contributions.

A powerful general method for treating end point contributions is provided by Mellin transforms (section 2.2). Key ideas are scaling transformations (section 2.3), disconnected scaling sets (section 2.3), independent scaling sets (section 2.3) and singular configurations (section 2.4). Multiple Mellin transforms (section 2.9) can be used to discuss limits in several variables.

Pinch contributions are discussed in section 2.6, where it is also explained how they can be evaluated by using end point techniques. An example of behaviour governed by a mixture of end points and pinches is given in section 2.7.

The treatment of divergences by dimensional regularisation is discussed in section 2.5 and the effect of divergences on asymptotic behaviour illustrated in section 3.3.

The determination of momentum flows associated with scaling

sequences is given in section 2.8, where the eikonal approximation is also worked out.

Sudakov parameters are defined in section 3.1. The importance of contour closing arguments in determining the significant range of values of Sudakov parameters in high energy regimes is illustrated in section 3.2. In section 3.3 a modified Sudakov representation with massless momenta is defined (see (3.3.13)) and in section 3.4 an alternative and universally powerful parametrisation for constituent momenta almost parallel to parent hadron momenta is written down ((3.4.3) *et seq.*).

A method by which Fourier transforms can be used to specify general analytic properties is given in section 3.6 (see (3.6.8)). The way $i\varepsilon$ prescriptions for internal invariants are specified by the $i\varepsilon$ prescriptions for external invariants is explained in section 4.2.

1
Feynman diagrams

1.1 Introduction

Relativistic quantum mechanics is a rich and intricate theory. It has defied general solution, so that its study has required recourse to models. These are chosen in the hope of reproducing the behaviour expected of the true theory in some appropriate extreme regime. Examples of such regimes are the Regge regime of high energy scattering at fixed momentum transfer (see Collins, 1977) and the deep inelastic scattering regime, in which the transfer of large momentum by weak or electromagnetic currents probes the constituents of hadronic matter (see Feynman, 1972). The results obtained from such models are of two kinds. Firstly the model may suggest that the scattering amplitude is constrained to take a restricted functional form in terms of the relevant variables. Examples drawn from the two regimes mentioned are, respectively, Regge pole behaviour $s^{\alpha(t)}$ (see section 2.1 *et seq.*) and the Bjorken scaling law $v^{-1}f(2v/-q^2)$ (see sections 2.10, 3.2). In fact more elaborate models suggest modifications to both these formulae (see sections 2.7 and 3.3) but the insight afforded by the simplest model is a valuable starting point for the discussion of each regime. Such kinematic results represent a stable component in our understanding of elementary particle physics. The second type of result sometimes extracted from the discussion of models is dynamic in character. It will try to predict the specific form of such functions as $\alpha(t)$ or $f(2v/-q^2)$ in terms of some precise physical mechanism. Clearly the successful accomplishment of this task is of the highest interest. However, in many regimes its attempt is somewhat ambitious in relation to current physical understanding and such results are liable to extensive revision as theory and experiment advance. The chief emphasis of this monograph will be on kinematic results of the first kind.

The oldest model used successfully to gain an insight into aspects of relativistic quantum mechanics is provided by the Feynman integrals of perturbation theory (see, for example, Bjorken & Drell, 1965). Interest in this model, and the methods associated with it, has recently been enhanced by the ideas of asymptotic freedom and the quantum chromodynamic theory of strong interactions (see Politzer, 1974). In such gauge theories the running coupling constant can be shown to become small like $(\ln v)^{-1}$ when all variables are large like v. Thus perturbation theory acquires a specific role in such theories at high energy.

Of course Feynman integrals only provide a model. However small the coupling constant may be there is no proof that the series is literally convergent. Nevertheless the perturbation series is formally a solution of the analytic and unitary properties which provide a basis for relativistic quantum mechanics (see Eden, Landshoff, Olive & Polkinghorne, 1966). Even here we must enter a note of caution. Feynman integrals have analytic and unitary properties expressed in terms of the particles associated with the propagators appearing in them. For quantum chromodynamics this means that they are expressed in terms of quarks and gluons. However the particles that figure in the S-matrix of the observable world are the hadrons. The relation between these two descriptions can only properly be understood when the bound state problem and confinement have been solved. Surely the eventual mastery of these problems will take us outside perturbation theory.

A partial way round this difficulty is afforded by the second model we are about to describe. Nevertheless Feynman integrals must retain their value as a guide to the fundamental interactions between quarks and gluons themselves (see, for example, the discussion of section 3.3). Their use is frankly heuristic but they have proved so powerful and convincing a guide to intuition that Feynman integrals have sometimes been described as a 'theoretical laboratory' in which conjectures on relativistic quantum mechanics can be tested. The second chapter of this monograph is intended as a handbook to the use of the apparatus of this laboratory.

The second main class of model of high energy processes stems from the work of Gribov (1968). Its earliest application was to Regge theory but later the method was used by Landshoff, Polkinghorne & Short (1971) to give a covariant formulation of the parton model of deep inelastic processes. Its nature can be illustrated by considering the process of fig. 1.1.1. The two thick external lines represent incident hadrons. They interact through the emission and scattering of constituents (or partons) corresponding to the lines 1 and 2 of the figure. The unshaded bubbles represent amplitudes for the emission of the partons, leaving a 'core' or residue behind. The shaded bubble corresponds to the scattering amplitude for the interaction of the two partons. These three bubbles are not to be thought of in a Feynman integral way at all. They are complete

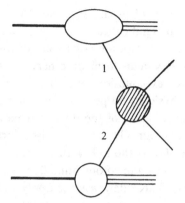

FIGURE 1.1.1 Constituent scattering.

(non-perturbative) subamplitudes for the subprocess (emission or scattering) which they represent. The kinematics of the overall process will constrain them to be evaluated in some specific regime. Many examples of this will be given in Chapter 3. Part of the specification of the model will include the stipulation of how the subamplitudes are to behave in their respective regimes. For example the shaded bubble might represent quark–quark scattering at high energy and large momentum transfer. It will be necessary then to have an *ansatz* for the behaviour of the amplitude in this regime. (Almost certainly this would be derived from applying perturbation theory analysis to quantum chromodynamics, so that the two models are closely interrelated.)

The Gribov approach is a hybrid one, for the subamplitudes are linked together by Feynman propagators associated with the interacting constituents. For example, the lines 1 and 2 of fig. 1.1.1 correspond to propagators of this type. Each high energy process modelled in this way will correspond to external momenta with large components in an appropriate frame of reference. The role of the propagators is to carry the flow of this large momentum through the interaction. The calculation of the effects of this flow is greatly facilitated by the use of Sudakov (1956) parameters. These give an expansion of all internal momenta in terms of the important external momenta. They thus facilitate a convenient covariant separation with large and small components. Full details of these techniques are given in chapter 3.

As far as possible the two main chapters, 2 and 3, are written in a self-contained way so that they are capable of being read independently of each other with the minimum of cross-reference. Chapter 2 may appear predominantly mathematical in character, for though it obtains many results of clear physical significance it does so by techniques which can be used to extract the asymptotic behaviour of functions defined by integrals, be they of Feynman form or otherwise. These methods are capable of systematic exploitation but their economic application – particularly to the physically important case of particles with spin – requires a clear understanding of the dominant pattern of momentum flow in the Feynman diagram. (This is emphasised particularly in sections 1.2 and 2.8.) This fact forms the link with the more obviously physical approach of chapter 3. When the momentum flow is properly understood the methods of chapter 3 are often easier to apply and they carry greater conviction because of their largely non-perturbative character.

Mueller (1970) has taught us how to use a generalised optical theorem to calculate cross-sections by taking discontinuities of scattering amplitudes in suitable variables. In simple cases such discontinuities are readily evaluated by elementary means. In chapter 4 we have collected together some more advanced discussion of discontinuities, making use of techniques drawn from both chapter 2 and chapter 3.

1.2 Feynman integrals and symmetric integration

An account of the origin of Feynman integrals in relativistic time-ordered perturbation theory can be found in introductory texts on quantum field theory (for example, Bjorken & Drell, 1965). In this section we shall consider only the simplest non-trivial example of a field theory, given by a real scalar field $\phi(x)$ of mass m interacting with itself through a term

$$g\,\phi^3(x)/3! \qquad (1.2.1)$$

in the Lagrangian. In the succeeding section we shall begin the consideration of more complicated and realistic theories involving particles with spin. These theories have additional complicating features not present in ϕ^3 theory but the latter provides a convenient illustration of many simple properties common to all expansions in terms of Feynman integrals.

The S-matrix for the scattering process may be written

$$S = 1 + iR, \qquad (1.2.2)$$

and the T-matrix is then defined in terms of the R-matrix by the equation

$$\langle p_{1f} \cdots p_{nf} | R | p_{1i} \cdots p_{mi} \rangle = (2\pi)^4 \delta(\textstyle\sum p_i - \sum p_f)\langle p_{1f} \cdots p_{nf} | T | p_{1i} \cdots p_{mi} \rangle.$$

$$(1.2.3)$$

The momenta $p_{1i} \ldots p_{mi}$ correspond to an initial state with m particles, and the momenta $p_{1f} \ldots p_{nf}$ to a final state with n particles. Each Feynman integral contributing to T can be associated with a Feynman diagram. Each diagram has external lines carrying the initial and final state momenta, there being one such line for each particle in the process. In addition the diagrams contain internal lines which correspond to the so-called virtual particles which mediate the interaction. Lines meet in trilinear vertices (corresponding to the trilinear interaction (1.2.1)). Internal lines must be connected to a vertex at both ends (since virtual particles are emitted and then reabsorbed) while external lines are only connected at one end (corresponding to the interaction in which the external particle is absorbed or emitted). These somewhat involved statements are illustrated by the examples of fig. 1.2.1 for the two particle–two particle scattering,

$$p_1 + p_4 \to p_2 + p_3, \tag{1.2.4}$$

(where we use the transparent notation of representing a particle by its four-momentum).

The momenta flowing in the internal lines are constrained by the requirement of four-momentum conservation at each vertex. The satisfaction of these constraint equations is analogous to a Kirchoff's law problem in electrical networks. The general solution is given in terms of l circulating loop momenta associated with the l independent loops of the diagram. For example in fig. 1.2.1a there is just one such loop with k the loop momentum, while in fig. 1.2.1b there are two independent loops with momenta k_1 and k_2 respectively.

Every topologically distinct diagram constructed in this way represents a contribution to $i(2\pi)^4 \delta(\sum p_i - \sum p_f) \langle T \rangle$ given by a Feynman integral specified by the following rules:

(i) a term $i/(2\pi)^4 \, 1/(q^2 - m^2 + i\varepsilon)$ for each internal line carrying momentum q. The infinitesimal $i\varepsilon$ is required to define how the pole at $q^2 = m^2$

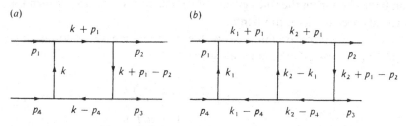

FIGURE 1.2.1 Typical Feynman diagrams.

is to be treated. The internal momenta q are linear combinations of the external momenta p and the loop momenta k, as in fig. 1.2.1.

(ii) a term $-\,i(\sum 2\pi)^4\,\delta^{(4)}(\sum q)$ for each vertex, where $\sum q$ is the algebraic sum of the momenta at the vertex. These δ-functions enforce energy–momentum conservation at each vertex and it is immediately possible to factor out from them the $(2\pi)^4\,\delta(\sum p_i - \sum p_f)$ which enforces overall energy–momentum conservation for the process.

(iii) an integral $\int d^4k$ for each independent loop momentum in the diagram.

(iv) a symmetry factor S^{-1} for the whole diagram, where S is the number of different ways in which the internal lines can be arranged with the external lines fixed. For example, S is 1 for the diagrams of fig. 1.2.1 but 2 for fig. 1.2.2 because of the possibility of interchanging the two internal lines.

FIGURE 1.2.2 A diagram with $S \neq 1$.

(In some books (e.g. Bjorken & Drell, 1965) the rules are given in a form which distributes the factors of 2π differently. The form given above is quite general and will hold for theories with trilinear or quadrilinear vertices. The other prescriptions are specific to the case of trilinear vertices.)

Symmetric integration

A diagram with l independent loops and n internal lines will have associated with it an integral which is proportional to

$$\lim_{\varepsilon \to 0} \int d^4k_1 \ldots d^4k_l / \prod_{r=1}^{n} (q_r^2 - m^2 + i\varepsilon). \qquad (1.2.5)$$

For convenience we have omitted writing the numerical factors which can be read off from the rules given above. The next task is to perform the integrals over the loop momenta.

An indispensable preliminary is the combination of denominators in (1.2.5) by using Feynman's identity

$$\frac{1}{d_1 d_2 \ldots d_n} = (n-1)! \int_0^1 \frac{d\alpha_1 \ldots d\alpha_n\,\delta(\sum \alpha - 1)}{\left[\sum_{i=1}^{n} \alpha_i d_i\right]^n}. \qquad (1.2.6)$$

In passing we may note the useful generalisation of (1.2.6)

$$\frac{1}{d_1^{1+r_1} d_2^{1+r_2} \ldots d_n^{1+r_n}} = \frac{(n + \sum r_i - 1)!}{\prod\limits_{i=1}^{n} r_i!} \int_0^1 \frac{\prod \alpha_i^{r_i} d\alpha_i \, \delta(\sum \alpha - 1)}{[\sum \alpha_i d_i]^{n + \Sigma r_i}}. \quad (1.2.7)$$

For integral r_i this can be proved from (1.2.6) by differentiating with respect to the d_i. It is also true for general r_i provided the factorials are interpreted via the gamma function: $n! \equiv \Gamma(n+1)$.

Although applying (1.2.6) to (1.2.5) appears to introduce a further complication in the form of an auxiliary Feynman parameter α_i associated with each internal line, we shall find that it enables the loop momenta k_j to be integrated out.

The expression (1.2.5) becomes

$$\Gamma(n) \frac{\int_0^1 \prod d\alpha_i \, \delta(\sum \alpha - 1) \prod \int d^4 k_j}{[\sum \alpha_i (q_i^2 - m^2) + i\varepsilon]^n}, \quad (1.2.8)$$

where there is now a single denominator. The qs are linear combinations of the loop momenta k and the external momenta p so that the denominator in (1.2.8) can be written as

$$\psi(p, k, \alpha) \equiv \sum \alpha_i (q_i^2 - m^2) + i\varepsilon$$
$$= k^T \cdot A \cdot k - 2k^T \cdot B \cdot p + (p^T \cdot \Gamma \cdot p - \sigma) + i\varepsilon, \quad (1.2.9)$$

where

$$\sigma = \sum \alpha_i m^2. \quad (1.2.10)$$

If there are l independent loop momenta and e external lines (and thus, by overall momentum conservation, $e - 1$ independent external momenta) A, B and Γ are respectively $l \times l$, $l \times (e - 1)$, and $(e - 1) \times (e - 1)$ matrices. Their elements are linear in the αs and the matrices act on row and column vectors made up of the l independent loop momenta (k) and the $e - 1$ independent external momenta (p). The superscript T represents the transpose in matrix space. In these row and column vectors each entry is a momentum and is thus itself a Lorentz vector. Thus the first term of (1.2.9) written out fully takes the form of a multiple sum over matrix and Lorentz indices

$$k^T \cdot A \cdot k = \sum_{i,j=1}^{l} \sum_{\mu=0}^{3} k_i^\mu \cdot A_{ij} \cdot k_{j\mu}, \quad (1.2.11)$$

and similarly for the other terms.

The next step is to remove the terms linear in k from (1.2.9) by the *displacement transformation*

$$k = k' + A^{-1} \cdot B \cdot p. \tag{1.2.12}$$

This is a legitimate change of variable in the infinite integrals over the $k_{j\mu}$ provided the integrals are convergent. However it is a well-known fact that it is often necessary to consider Feynman integrals which are formally infinite and which have to be regularised and subject to the renormalisation procedure (see Bjorken & Drell, 1965; and section 2.5 below). A formal manipulation like (1.2.12) can still be justified if the k-integrals are only logarithmically divergent. However, if the integral is linearly divergent, an extra term appears on making the transformation (1.2.12). This is usually called the 'surface term' since it can be associated with the integral over a complete divergence, which in the linearly divergent case gives on integration a non-vanishing contribution from the sphere at infinity. (See appendix A5 of Jauch & Rohrlich (1955) for details.) These facts are illustrated by the paradigm identities

$$\int \frac{\mathrm{d}^4 k}{[(k-p)^2 - a^2]^2} = \int \frac{\mathrm{d}^4 k'}{[k'^2 - a^2]^2}, \tag{1.2.13a}$$

$$\int \frac{k_\mu \mathrm{d}^4 k}{[(k-p)^2 - a^2]^2} = \int \frac{(k'_\mu + p_\mu)\mathrm{d}^4 k'}{[k'^2 - a^2]^2} - \frac{\mathrm{i}\pi^2}{2} p_\mu, \tag{1.2.13b}$$

the second term in (1.2.13b) being the surface term.

The next step following (1.2.12) is to diagonalise the quadratic term $k'^\mathrm{T} \cdot A \cdot k'$ in matrix space by an orthogonal transformation,

$$k' = R \cdot k'', \quad R^\mathrm{T} \cdot R = 1, \quad R^\mathrm{T} \cdot A \cdot R = A'' \text{ (diagonal)}, \tag{1.2.14}$$

so that ψ now becomes

$$k''^\mathrm{T} \cdot A'' \cdot k'' - (B \cdot p)^\mathrm{T} \cdot A^{-1} \cdot (B \cdot p) + p^\mathrm{T} \cdot \Gamma \cdot p - \sigma. \tag{1.2.15}$$

Since R is orthogonal its determinant is unity and the Jacobian of the transformation from k to k'' is therefore 1:

$$\prod \int \mathrm{d}^4 k_j \to \prod \int \mathrm{d}^4 k''_j. \tag{1.2.16}$$

The k'' are Lorentz vectors so that k''^2 is of indefinite sign. In order to perform the integrations it is necessary to transform to an integral with a definite metric for the k''. This can be achieved by rotating the contours of integration for the timelike component of each k'' so that it runs along the imaginary axis:

$$k''_{j0} \rightarrow i\,k''_{j0}. \tag{1.2.17}$$

The loop momenta now become effectively Euclidean vectors with $k''^2 \leqslant 0$. (There are non-trivial problems in justifying this. The integrand (1.2.8) has poles which are displaced from the contour of integration by the iε prescription. The sense of these displacements must be such as to permit the anticlockwise contour rotation implied by (1.2.17). For a denominator like that of a simple propagator,

$$k''^2 - a^2 + i\varepsilon, \tag{1.2.18}$$

with $a^2 > 0$, the Feynman iε prescription produces the contour of fig. 1.2.3 which avoids the poles at $k''_0 = \pm\sqrt{(k^2 + a^2)}$ in just the right sense for the rotation to be possible. However with the more complicated denominator structure of (1.2.8) and (1.2.9) this is not manifestly the case. The answer lies in first evaluating the integral in an unphysical region where the

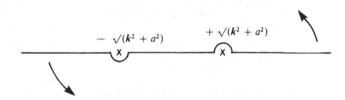

FIGURE 1.2.3 The Feynman contour in the k''_0-plane. Arrows show how the contour can be rotated.

external vectors are also Euclidean (rather than Lorentz). In this region the denominator takes the form (1.2.18) for each loop momentum with $a^2 \geqslant 0$, so that the rotation is justified. It will then be necessary to obtain the integral in the physical region by analytic continuation from the unphysical region. We discuss this continuation below. An alternative approach makes use of the exponential representation (1.2.33).)

Once the loop momenta have been made Euclidean by the transformation (1.2.11) it is straightforward to integrate out each component in turn by using repeatedly the identity

$$\int_{-\infty}^{\infty} \frac{dk''_{j\mu}}{(-a''_j k''^2_{j\mu} - b)^n} = (-1)^n \frac{\Gamma(\tfrac{1}{2})\Gamma(n - \tfrac{1}{2})}{\Gamma(n)} \frac{1}{b^{n-1/2} a''^{1/2}_j}. \tag{1.2.19}$$

Each application of this formula reduces the power of the residual denominator (b) by $\tfrac{1}{2}$ and introduces a factor of $(a''_j)^{-1/2}$, where a''_j is the coefficient of $-k_{j\mu}^2$ in the diagonalised quadratic form of (1.2.14), (1.2.15). The minus sign arises from the Euclidean metric with $k''^2 \leqslant 0$. Thus the total effect of performing the $4 \times l$-fold integration over loop momentum components is equivalent to replacing the denominator ψ of (1.2.15) by its value with $k'' = 0$, reducing the power of ψ in (1.2.8) from n to $n - 2l$, and introducing a factor of

$$\prod_j \frac{1}{(a_j'')^2} = \frac{1}{(\det A'')^2} = \frac{1}{(\det A)^2}, \qquad (1.2.20)$$

together with the numerical factors resulting from the successive use of (1.2.19). In (1.2.20) we have used the fact that the product of the eigenvalues of a symmetric matrix is equal to its determinant and the determinant is unaltered by orthogonal transformation. Thus symmetric integration yields a result proportional to

$$\Gamma(n - 2l) \int \frac{\prod d\alpha_i \delta(\sum \alpha - 1) [C(\alpha)]^{n - 2l - 2}}{[D(\alpha, p) + i\varepsilon C(\alpha)]^{n - 2l}}, \qquad (1.2.21)$$

where

$$C(\alpha) = \det A,$$

$$D(\alpha, p) = -(B \cdot p)^{\mathrm{T}} \cdot X \cdot (B \cdot p) + (p^{\mathrm{T}} \cdot \Gamma \cdot p - \sigma)C, \qquad (1.2.22)$$

with

$$\sigma = \sum \alpha_i m^2, \quad X = A^{-1} C = \operatorname{adj} A. \qquad (1.2.23)$$

An alternative definition of D is given by

$$D = C\psi_0, \qquad (1.2.24)$$

where ψ_0 is obtained from ψ by eliminating the original loop momenta k_j using the equations

$$\partial \psi / \partial k_j = 0, \quad \text{each } j. \qquad (1.2.25)$$

These equations are just

$$A \cdot k = B \cdot p, \qquad (1.2.26)$$

that is, (1.2.12) with k' set equal to zero. Note that the dimensionality of space enters (1.2.21) in two ways: the 2 multiplying l in the exponents of C and D and the 2 which is the difference between the two exponents. If space were n-dimensional both these 2's would be replaced by $\frac{1}{2}n$.

This perhaps rather odd observation will prove to be of relevance later (see section 2.5).

Analyticity

The functions defined by Feynman integrals have interesting analytic properties which have been extensively studied. (See Eden *et al.*, 1966, ch. 2.) The natural variables to consider are invariants formed from scalar products of the external momenta. For example, for the two-particle scattering diagrams of fig. 1.2.1 the obvious choice is the familiar Mandelstam set

$$\left.\begin{array}{l} s = (p_1 + p_4)^2 = (p_2 + p_3)^2, \\ t = (p_1 - p_2)^2 = (p_3 - p_4)^2, \\ u = (p_1 - p_3)^2 = (p_2 - p_4)^2, \end{array}\right\} \qquad (1.2.27)$$

together with the external masses p_i^2, to which they are related by the equation

$$s + t + u = \sum p_i^2. \qquad (1.2.28)$$

While in physical applications the external masses are fixed at their physical values it is sometimes necessary to consider them as analytic variables. For example, continuation to an unphysical region where all the external vectors are Euclidean will involve continuation in the p_i^2.

The simplest singularities are the *normal thresholds* corresponding to the opening up of new channels. In a variable like s they occur at $s = (nm)^2$, $n = 2, 3 \ldots$. If we think of s as the square of the centre-of-mass energy for the process $p_1 + p_4 \to p_2 + p_3$ then these singularities are the threshold values of s at which $2, 3, \ldots$ particle states first become possible for that channel. There will be similar singularities in the variables t and u.

The role of the $i\varepsilon$ in (1.2.5), (1.2.21) is to make the integral well defined even in regions where the denominators may vanish. In analytic terms this question of definition is just the question of choosing the right Riemann sheet of the many-valued function, with all its branch points, which is defined by the Feynman integral. The way the $i\varepsilon$ prescription does this is particularly transparent. We can interpret (1.2.5) as assigning a negative imaginary part to all internal masses m^2 while keeping all other variables real. Thus this will, for example, keep s real but depress the normal thresholds $(nm)^2$ just below the real axis. Reinterpreting this infinitesimal displacement we can equivalently say that the correct physical prescription is to keep variables like s just *above* their normal threshold cuts.

Exponential representations

It is sometimes convenient to write (1.2.21) in the alternative form

$$(-1)^n \prod \int_0^\infty d\tilde{\alpha}_i\, C(\tilde{\alpha})^{-2} \exp\left[D(\tilde{\alpha},p)/C(\tilde{\alpha})\right]. \qquad (1.2.29)$$

The $\tilde{\alpha}_i$ correspond to the α_i but are no longer constrained to have sum unity. $C(\tilde{\alpha})$ turns out to be homogeneous of degree l in the $\tilde{\alpha}$ and positive definite for positive $\tilde{\alpha}$s (see below). $D(\tilde{\alpha},p)$ is homogeneous of degree $(l+1)$ in the $\tilde{\alpha}$s. In order that (1.2.29) is well defined we must have D negative definitive. This will be so if we are in the unphysical region corresponding to Euclidean vectors. Then there is no need for an explicit $i\varepsilon$ in (1.2.21) since the denominator does not vanish.

The equivalence of (1.2.21) and (1.2.29) in the Euclidean region can be shown by the scaling transformation:

$$\tilde{\alpha}_i = \rho\alpha_i, \quad \sum\alpha_i = 1,$$

$$\prod \int_0^\infty d\tilde{\alpha}_i \to \int_0^\infty \rho^{n-1} d\rho \prod \int_0^1 d\alpha_i \delta(\sum\alpha - 1). \qquad (1.2.30)$$

The ρ-integration can be performed explicitly by using the identity

$$\int_0^\infty \rho^{n-1} \rho^{-2l} \exp\left[-\rho(-D/C)\right] d\rho = \Gamma(n-2l)(-C/D)^{n-2l}, \quad -D/C > 0, \qquad (1.2.31)$$

and the equivalence follows.

Alternatively (1.2.29) could be derived directly by using an exponential representation for the propagator

$$\frac{1}{q^2 - m^2} = -\int_0^\infty d\tilde{\alpha} \exp\left[\tilde{\alpha}(q^2 - m^2)\right], \qquad (1.2.32)$$

which is valid if $q^2 \leqslant 0$. Employment of such an expression for each propagator in a Feynman integral will yield an integral with the quadratic form ψ in the exponential. Symmetric integration techniques will then lead to Gaussian integrals over the components of the k'' loop momenta. These integrals can then readily be evaluated to give (1.2.29).

It is sometimes convenient to replace (1.2.32) by the complex exponential

$$\frac{1}{q^2 - m^2 + i\varepsilon} = -i \int_0^\infty d\tilde{\alpha} \exp\left[i\tilde{\alpha}(q^2 - m^2 + i\varepsilon)\right]. \qquad (1.2.33)$$

This has the advantage of being defined (via the $i\varepsilon$) for all real q^2. Its use will yield an expression similar to (1.2.29) but with exponent i D/C, in which the $i\varepsilon$ in D now plays an essential role. An advantage of such complex exponential representation is that the loop momenta integrations over the k'' can be performed for Lorentz vectors without the need to make the transformation (1.2.17) to the Euclidean region.

C- and D-functions

Symmetric integration is a straightforward procedure but its detailed application is cumbersome for all but the simplest diagrams. Fortunately however simple rules are known which enable the C- and D-functions to be written down directly from the topological structure of the diagrams. These rules are best illustrated by reference to a particular case, for which we choose fig. 1.2.4. The numbers indicate the labelling of the lines (and hence the Feynman parameters) of the diagram.

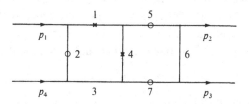

FIGURE 1.2.4 A Feynman diagram with crosses representing a cut contributing to C and circles a cut contributing to the coefficient of $(p_1 + p_4)^2$ in D.

The rules for writing down the function $C(\alpha)$ are as follows.

Lines are cut in such a way that (i) every vertex is connected to every other by a sequence of lines still uncut; (ii) no further cuts can be made without violating (i). The lines marked with crosses in fig. 1.2.4 represent a pair of cut lines satisfying these criteria. (The number of lines it is possible to cut in this way always equals l, the number of independent loops.) The product of the αs corresponding to the cut lines ($\alpha_1 \alpha_4$ in our example) is one term appearing in C. The function $C(\alpha)$ is the sum of all such products corresponding to all permissible sets of cuts lines. For fig. 1.2.4 this gives

$$C(\alpha) = \alpha_1(\alpha_4 + \alpha_5 + \alpha_6 + \alpha_7) + \alpha_2(\alpha_4 + \alpha_5 + \alpha_6 + \alpha_7)$$
$$+ \alpha_3(\alpha_4 + \alpha_5 + \alpha_6 + \alpha_7) + \alpha_4(\alpha_5 + \alpha_6 + \alpha_7)$$
$$= (\alpha_1 + \alpha_2 + \alpha_3 + \alpha_4)(\alpha_4 + \alpha_5 + \alpha_6 + \alpha_7) - \alpha_4^2. \quad (1.2.34)$$

Notice that by construction C is positive definite for positive definite αs. It is also easy to see that C will vanish if all the αs round a closed loop are put equal to zero, for each term in C must contain at least one α from each loop.

The part of D which depends upon the internal masses m is immediately calculable from (1.2.23), once C is known. We now give the rules for calculating the part of D which contains the dependence upon the external momenta. It will be a linear function of the invariants constructed out of the external momenta, and the coefficients multiplying these invariants are determined by the following prescription.

By means of a set of cuts on internal lines divide the diagram into two disjoint halves, each satisfying the rules (i) and (ii) given above and each having at least one external momentum attached to it. Each such set of cuts yields a term in D given by the product of the αs of the cut lines times the invariant $(\sum'p)^2$, where $\sum'p$ is the algebraic sum of the external momenta attached to one of the halves. (Momentum conservation makes the choice of which half irrelevant.) Thus the lines marked with circles in fig. 1.2.4 represent cuts made in accordance with these rules, which divide the diagram into two parts with momentum $p_1 + p_4 (= p_2 + p_3)$ fed into one (or other) half. There is therefore a term

$$\alpha_2\alpha_5\alpha_7(p_1 + p_4)^2 \quad (1.2.35)$$

in D. The complete part of D for fig. 1.2.4 dependent on external momenta is given by

$$(p_1 + p_4)^2 \left[\alpha_1\alpha_3(\alpha_4 + \alpha_5 + \alpha_6 + \alpha_7) + \alpha_5\alpha_7(\alpha_1 + \alpha_2 + \alpha_3 + \alpha_4)\right.$$
$$\left. + \alpha_1\alpha_4\alpha_7 + \alpha_3\alpha_4\alpha_5\right]$$
$$+ (p_1 - p_2)^2 \left[\alpha_2\alpha_4\alpha_6\right] + p_1^2\left[\alpha_1\alpha_2(\alpha_4 + \alpha_5 + \alpha_6 + \alpha_7) + \alpha_2\alpha_4\alpha_5\right]$$
$$+ p_2^2\left[\alpha_5\alpha_6(\alpha_1 + \alpha_2 + \alpha_3 + \alpha_4) + \alpha_1\alpha_4\alpha_6\right]$$
$$+ p_3^2\left[\alpha_6\alpha_7(\alpha_1 + \alpha_2 + \alpha_3 + \alpha_4) + \alpha_3\alpha_4\alpha_6\right]$$
$$+ p_4^2\left[\alpha_2\alpha_3(\alpha_4 + \alpha_5 + \alpha_6 + \alpha_7) + \alpha_2\alpha_4\alpha_7\right]. \quad (1.2.36)$$

It will be clear that D is homogeneous of degree $l + 1$ for a diagram with l independent loops and that it vanishes when all the αs round a loop are put equal to zero.

The coefficients of all the $(\sum'p)^2$ are by construction positive definite for positive αs. However, not all the $(\sum' p)^2$ are independent variables and the coefficients of the *independent* variables may be of indefinite sign for positive αs. It is only possible to cut fig. 1.2.4 to give coefficients of $s = (p_1 + p_4)^2$ and $t = (p_1 - p_2)^2$ and there is no way in which it can be cut according to the rules (i) and (ii) to give a term that would be a coefficient of $u = (p_1 - p_3)^2$. Thus only the independent variables s and t appear in (1.2.34). However this is not so for a diagram like fig. 1.2.5. In this case D contains all three Mandelstam variables with coefficients given by

$$s\left[\alpha_2\alpha_3(\alpha_4 + \alpha_5 + \alpha_6 + \alpha_7) + \alpha_3\alpha_4\alpha_7 + \alpha_2\alpha_5\alpha_6\right] + t\left[\alpha_1\alpha_6\alpha_7\right]$$
$$+ u\left[\alpha_1\alpha_4\alpha_5\right]. \tag{1.2.37}$$

If (1.2.28) is used to express D in terms of the two independent variables s and t, then their coefficients will involve products of αs with both signs.

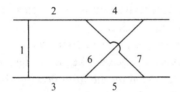

FIGURE 1.2.5 A non-planar diagram.

Clearly the difference between fig. 1.2.4 and fig. 1.2.5 is that the former can be drawn on a plane piece of paper with lines meeting only at interaction vertices, while the latter can not be drawn on such a piece of paper without a pair of lines crossing which do not interact however one tries to permute the external lines in an attempt to avoid this. Diagrams of the first type are called *planar* and those of the second type *non-planar*. The distinction is one that we shall find to be of importance.

Momentum flow

The momentum q_i in the ith line of a diagram can be written as a sum of the appropriate symmetric loop momenta k' (or equivalently k'') together with a linear combination of the external momenta p. The latter term, which we shall denote by Y_i, arises from the explicit external momenta which may be associated with the line in the original labelling of the momenta (see fig. 1.2.1 for example) together with the further displacement term from (1.2.12). The combination of these two gives a Y_i which is intrinsic to the diagram and independent of the original choice of how to define the loop momenta k. Again simple rules can be given for writing down the Y_i from the topological structure of the diagram:

Choose those cuts in the calculation of D given above which include the ith line. Then there is a term in Y_i which consists of $(\sum' p) C^{-1}$ multiplied by the αs of the cut lines other than i itself. Y_i is the sum of all such terms. It is necessary to define carefully the sense of $\sum' p$ for the prescription. Considering the subdiagram to which these external momenta are attached, if q_i is directed out of (into) this subdiagram then $\sum' p$ is directed into (out of) the subdiagram.

For example in fig. 1.2.4 we find

$$Y_1 = C^{-1}\big[p_1\alpha_2(\alpha_4 + \alpha_5 + \alpha_6 + \alpha_7)$$
$$+ (p_1 + p_4)(\alpha_3(\alpha_4 + \alpha_5 + \alpha_6 + \alpha_7) + \alpha_4\alpha_7) + p_2\alpha_4\alpha_6\big].$$

(1.2.38)

A proof of these rules is easily given by using the trick of auxiliary momenta described in the following section.

The Y_i often have a direct physical significance. Suppose that one is dealing with a high energy process, like those described in section 1.1, in which there is large external momentum flowing through the diagram. Suppose moreover that the dominant behaviour of the Feynman integral is known to come from a region of integration in which the symmetric loop momenta k'' are small compared with the scale set by the external momenta. Then the Y_i will specify the way in which the large momentum flows through the diagram.

1.3 Spin

For the discussion of realistic theories it is necessary to extend the treatment of the previous section to include particles with spin. This leads to the presence of extra numerator factors associated with the propagators of internal lines:

$(\gamma \cdot q + m)$ for a particle of spin $\tfrac{1}{2}$;

$(q_{\mu\nu} - \theta q_\mu q_\nu / q^2)$ for a particle of spin 1, (θ being a gauge-dependent parameter).

Further numerator factors appear associated with vertices, composed of γ-matrices and/or further powers of momenta q_μ, and there may also be quadrilinear vertices corresponding to the presence of four-field terms in the Lagrangian (see the rules in Bjorken & Drell, 1965). For our present purpose we concentrate on the consequences of the appearance of extra powers of momenta in the numerator. These produce two effects:

(i) Scalar products of external momenta are formed from scalar products of the Y_i.

(ii) The presence of extra powers of k' affects the results of symmetric integration. One can show, for example, (Chisholm, 1952) that the presence of the single term $k'_{j_1} \cdot k'_{j_2}$ (where j_1 and j_2 label the loops from which the loop momenta k' are drawn) produces an extra factor of

$$[\text{adj } A]_{j_1 j_2}/C \tag{1.3.1}$$

in (1.2.21).

Patient application of symmetric integration techniques to these more complicated cases will in principle determine the effect of numerator factors in any Feynman integral. However, the procedure is laborious to apply in all but the simplest instances, though often the work can be simplified in particular applications where only some of the many terms may be of dominating significance. This latter aspect will be illustrated by some of the investigations of chapter 2. No useful set of rules, comparable to those for C and D, are known for handling numerator factors. However, the following trick is sometimes of value when the number of such factors is small.

Auxiliary momenta

Suppose that the momentum $q_{i\mu}$, associated with the ith line, appears in the numerator. Introduce a dummy auxiliary momentum a_μ which is fed into the diagram so that it only flows through the ith line. We can conceive of this as entering the line at one end either through a real external line (if there is one present) in which the momentum is now $p + a$ rather than p, or through a fictitious external line carrying a (if there is no genuine external line present). The momentum a flows out at the other end of the ith line in an analogous fashion. These possibilities are illustrated in fig. 1.3.1.

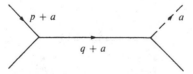

FIGURE 1.3.1 A dummy momentum a fed in through a real external line and out through a fictitious external line.

We can use the trivial identity

$$\frac{q_\mu + a_\mu}{(q+a)^2 - m^2} = -\int_0^\infty d\tilde{\alpha}\left(\frac{1}{2\tilde{\alpha}}\frac{\partial}{\partial a_\mu}\right)\exp\left[\tilde{\alpha}(q+a)^2 - m^2\right] \qquad (1.3.2)$$

in place of (1.2.31) and put $a_\mu = 0$ at the end of the calculation to recover $q_\mu/(q^2 - m^2)$. If there are different numerator factors $q_{i\mu}$ associated with different lines then auxiliary momenta $a_{i\mu}$ will have to be fed in and out of each such line separately. If more than one power of the same momentum appears in the numerator it is necessary to bear in mind that a_μ is not set equal to zero until all differentiations have been performed. Therefore we must use formulae like

$$\frac{(q_\mu + a_\mu)(q_\nu + a_\nu)}{(q+a)^2 - m^2} = -\int_0^\infty d\tilde{\alpha}\left[\left(\frac{1}{2\tilde{\alpha}}\frac{\partial}{\partial a_\mu}\right)\left(\frac{1}{2\tilde{\alpha}}\frac{\partial}{\partial a_\nu}\right) - \frac{g_{\mu\nu}}{2\tilde{\alpha}}\right]$$
$$\times \exp\left[\tilde{\alpha}(q+a)^2 - m^2\right]. \qquad (1.3.3)$$

By the use of identities like (1.3.2) and (1.3.3) the original integral with numerator factors can be written in the form of a differential operator $\mathcal{D}_{a_i}$ in the auxiliary momenta a_i, acting on an integral in which only simple propagator factors (1.2.31) appear without additional numerator terms. The result of symmetric integration on such an integral is just to give C- and D-functions specified by the rules of the last section, provided that it is remembered that D is a function of both ps and as, treated together as external variables, and the existence of the (fictitious) external lines required for the as must consequently be taken into account in applying the rules for D. The explicit dependence of D on the as which is now exhibited permits the effect of the differential operator $\mathcal{D}_{a_i}$ to be evaluated

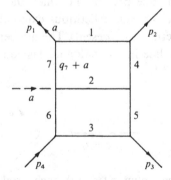

FIGURE 1.3.2 A diagram with a dummy variable a inserted.

and finally the as can be set equal to zero to evaluate the original expression.

As a simple example suppose we consider the diagram of fig. 1.3.2 with a single numerator factor $q_{7\mu}$ for which an auxiliary variable a must be inserted as shown. Many a-dependent terms will appear in D, one of which is, for example,

$$(p_4 + a)^2 \alpha_7 \alpha_2 \alpha_3. \qquad (1.3.4)$$

On differentiation this will produce an additional factor in (1.2.21) of the form

$$(\alpha_2 \alpha_3 / C(\alpha)) p_{4\mu}.$$

Similarly the coefficient of $(p_1 + a)^2$ will give a term proportional to $p_{1\mu}$, and so on. It is easy to see that the α_7^{-1} factor in the a-differential operator always cancels out.

This technique of auxiliary momenta can be used to derive (1.3.1) and also to prove the rules for Y_i given at the end of the last section.

2
Asymptotic behaviour in perturbation theory

2.1 Introduction

The regime of interactions at high energy s and fixed momentum transfer t is the subject of Regge theory. For a full description of the idea of complex angular momentum l and its application to particle physics, reference should be made to suitable texts (for example, Collins, 1977). The important results for the purpose of model making relate to the asymptotic behaviour of scattering amplitudes. Thus a Regge pole singularity at

$$l = \alpha(t) \qquad (2.1.1)$$

will give a high energy behaviour proportional to

$$s^{\alpha(t)}. \qquad (2.1.2)$$

The trajectory function $\alpha(t)$ has a double physical significance. When t is a momentum transfer and so less than or equal to 0, $\alpha(t)$ governs the high energy behaviour in the s-channel, as indicated by (2.1.2). However, when t is continued to positive values, so that it can be interpreted as an energy in the crossed channel, then the values of t which make $\alpha(t)$ pass through a non-negative integer n correspond to bound states or resonances of angular momentum n in that crossed channel and the appropriate poles are present in the scattering amplitude.

In addition to Regge poles, giving a pure power behaviour in s, there are Regge cut singularities giving a behaviour of the form

$$\int^{\alpha_c(t)} d\alpha' \, \rho(\alpha') s^{\alpha'}, \qquad (2.1.3)$$

and corresponding to a branch point at $l = \alpha_c(t)$. The use of models has played a particularly important role in demonstrating the existence of Regge cuts and exhibiting their properties.

Regge theory supposes the existence of Regge pole singularities. The question of whether this is reasonable can only be answered by interrogating suitable models. At first sight perturbation theory would not seem a likely candidate since the bound states associated with Regge

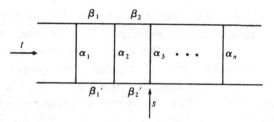

FIGURE 2.1.1 A ladder with n rungs.

poles are clearly a non-perturbative phenomenon. The answer lies in considering infinite sums of Feynman diagrams, such as the set of ladders illustrated by fig. 2.1.1. Such an infinite set allows the two particles represented by the sides of the ladder to interact an unlimited number of times by exchanging the virtual particles represented by the rungs of the ladder. Physically we can see that such unlimited interaction is just what one needs to give the possibility of a bound system. Formally the point is made by noting that the set of ladders provides a solution of the relativistic bound state equation of Bethe & Salpeter (1951) in the one-particle exchange approximation. Our first task therefore will be to evaluate the leading asymptotic behaviour at large s and fixed t of the Feynman integral associated with each ladder and then to sum the resulting terms. We shall find that a behaviour of the form of (2.1.2) is indeed found. Of course the procedure is frankly heuristic. There is no mathematical necessity for the sum of the leading behaviours of the terms of the series to give the leading behaviour of the sum, even if we assume this latter to exist. However the insight afforded by the procedure is compelling.

Asymptotic behaviour of integrals

Before turning to this task it will be convenient to consider a few simple mathematical facts about the asymptotic behaviour of integrals, which will illustrate the techniques that we shall use extensively in what follows.
    Consider

$$F(s) \equiv \int_a^b \frac{\mathrm{d}x}{(sx+d)^2} = \frac{1}{s}\left(\frac{1}{as+d} - \frac{1}{bs+d}\right), \qquad (2.1.4)$$

as $s \to \infty$ with a, b, d fixed. Inspecting the integral one might expect that the asymptotic behaviour of $F(s)$ is s^{-2} since that is the behaviour of the integrand. If $a, b \neq 0$ this is indeed the case. We call such behaviour, read off from the integrand, *natural behaviour*. However if either a or b

is zero then $F(s)$ has the enhanced asymptotic behaviour s^{-1}. This is clearly due to the fact that the coefficient of s in the denominator of the integrand vanishes at $x = 0$ so that the s^{-2} argument breaks down at this point. Notice however that this only affects the asymptotic behaviour if $x = 0$ is an *end point* of integration. If $a < 0$, $b > 0$, so that $x = 0$ lies within the region of integration, the asymptotic behaviour is unmodified. The reason for this is that, if we think of the x-integral in complex variable terms, Cauchy's theorem allows one to modify the interior of the region of integration by distorting the contour away from $x = 0$. If $x = 0$ is an end point this is not possible. Only in this latter case is the behaviour of the integrand at $x = 0$ of inescapable significance for the asymptotic behaviour of the integral.

In section 2.6 we shall encounter a further example in which distortion from $x = 0$ is prevented even in the interior of the region of integration by the presence of singularities which trap the contour at $x = 0$. Such contributions are called *pinches*.

Another way of thinking about the enhancement that an end point gives over natural behaviour is as follows. The natural behaviour of the integral (2.1.4) can be written

$$s^{-2} \int_a^b \frac{\mathrm{d}x}{x^2}, \qquad (2.1.5)$$

and if $a = 0$ the divergence of the integral in (2.1.5) at its lower end point signals that the asymptotic behaviour has been incorrectly estimated. In fact the signal is a rather precise one, for the linear divergence of the integral at $x = 0$ corresponds to the fact that a further linear power s is needed to give the correct asymptotic behaviour. This is very characteristic of asymptotic calculations. The presence of a divergent integral in a coefficient indicates an error in the estimate, the part of the region of integration giving the divergence pinpoints the source of the true asymptotic behaviour, and the degree of the divergence indicates its effect.

Consider now Feynman integrals in the Regge regime of large s and fixed t. Since D is linear in the external invariants the integral is proportional to

$$\int_0^1 \prod \mathrm{d}\alpha_i \, \frac{\delta(\sum \alpha - 1) [C(\alpha)]^{m-2}}{[g(\alpha)s + d(t,\alpha)]^m}. \qquad (2.1.6)$$

Here we are considering the simple case of spinless scalar particles where all the s-dependence is exhibited in the denominator. When particles with spin are present the numerator becomes more complicated and may contain powers of s. We consider such theories in section 2.8.

The insight afforded by the example (2.1.4) suggests that the dominant asymptotic behaviour of (2.1.6) will come from the neighbourhood of points where $g(\alpha) = 0$, provided that this vanishing is inescapable in the sense explained above. This can be due either to its being an end point (some $\alpha_i = 0$) or to the pinch phenomenon to be treated in section 2.6. For planar diagrams g is a sum of products of αs with positive signs and can only vanish at an end point.

Finally we note a general formula which is of considerable utility in evaluating end point contributions

$$\int_0^\varepsilon \rho_1^{m-1} d\rho_1 \cdots \int_0^\varepsilon \rho_n^{m-1} d\rho_n \frac{1}{(\rho_1 \rho_2 \cdots \rho_n \bar{g} s + d)^p}$$

$$\sim \frac{(m-1)!(p-m-1)!}{(p-1)!} \frac{1}{\bar{g}^m d^{p-m}} s^{-m} \frac{(\ln s)^{n-1}}{(n-1)!}, \quad s \to \infty, \ p > m. \quad (2.1.7)$$

If $p = m$ the leading behaviour is proportional to $s^{-m}(\ln s)^n$. If $p < m$ the integral for the coefficient of the natural behaviour s^{-p} is convergent and this is the true asymptotic behaviour in this case.

Ladder diagrams

A simple example is provided by the ladder diagram of fig. 2.1.1 (Federbush & Grisaru, 1963; Polkinghorne 1963a). The associated integral is

$$g^2 \left(\frac{-g^2}{16\pi^2} \right)^{n-1} (n-1)! \int \frac{\prod d\alpha \prod d\beta \delta(\sum \alpha + \sum \beta - 1) [C(\alpha,\beta)]^{n-2}}{[\alpha_1 \alpha_2 \cdots \alpha_n s + d(t,\alpha,\beta)]^n}, \quad (2.1.8)$$

where d is simply given by

$$d = D\big|_{s=0}. \quad (2.1.9)$$

The coefficients of s vanishes when any of the parameters associated with the rungs of the ladder vanish:

$$\alpha_i = 0. \quad (2.1.10)$$

To evaluate the leading asymptotic behaviour it is necessary only to integrate in the neighbourhood of the points (2.1.10), that is, to consider the integrals

$$\prod_i \int_0^\varepsilon d\alpha_i, \quad (2.1.11)$$

It will then be sufficient to put $\alpha_i = 0$ elsewhere in the integrand in the leading approximation. That is, C will be replaced by

$$c(\beta) = C(0,\beta), \tag{2.1.12}$$

and d by

$$\delta(t,\beta) = d(t,0,\beta). \tag{2.1.13}$$

The leading asymptotic behaviour can be evaluated by using a particular case of the convenient formula (2.1.7). This leads to the expression

$$g^2 s^{-1} \frac{(\ln s)^{n-1}}{(n-1)!} \left[\left(\frac{-g^2}{16\pi^2} \right)^{n-1} (n-2)! \int \prod d\beta \frac{\delta(\sum \beta - 1)(c(\beta))^{n-2}}{(\delta(t,\beta))^{n-1}} \right]. \tag{2.1.14}$$

The functions $c(\beta)$ and $\delta(t,\beta)$ are the Feynman C- and D-functions which would be associated with the contracted diagram fig. 2.1.2 formed by short-circuiting the lines (rungs) for which $\alpha_i = 0$. In fact the expression in square brackets in (2.1.14) is exactly the Feynman integral corresponding to fig. 2.1.2 evaluated, however, with two-dimensional rather than four-dimensional loop momenta. This latter fact is indicated by the exponents of c and δ in (2.1.14) differing only by one (see section 1.2).

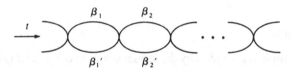

FIGURE 2.1.2 The diagram obtained by contracting the rungs of fig. 2.1.1.

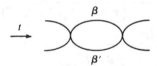

FIGURE 2.1.3 The basic bubble iterated in fig. 2.1.2.

It means, of course, that the diagram fig. 2.1.2, which would be divergent in four dimensions, gives in this two-dimensional case a well-defined function of t. The 'hinged' structure of the diagram – that is, its character as an $(n-1)$-fold iteration of the basic bubble fig. 2.1.3 – means that the quantity in square brackets in (2.1.14) can be written as

$$[K(t)]^{n-1}, \tag{2.1.15}$$

where $K(t)$ is the integral associated with fig. 2.1.3 in two dimensions and is therefore given by

$$K(t) = \frac{-g^2}{16\pi^2} \int_0^1 \frac{\mathrm{d}\beta \mathrm{d}\beta' \delta(\beta + \beta' - 1)}{(\beta\beta' t - m^2)} . \qquad (2.1.16)$$

Thus the leading asymptotic behaviour of the n-runged ladder is found to be

$$g^2 s^{-1} \frac{(\ln s)^{n-1}}{(n-1)!} [K(t)]^{n-1}, \qquad (2.1.17)$$

which if summed over n yields

$$g^2 s^{-1 + K(t)}. \qquad (2.1.18)$$

This is just Regge pole behaviour (2.1.2) with the trajectory function given to this approximation by

$$\alpha(t) = -1 + K(t). \qquad (2.1.19)$$

We have derived this encouraging result by the heuristic procedure of summing the asymptotic behaviour of individual Feynman diagrams evaluated in the 'leading logarithm' approximation. The same result was first obtained by Lee & Sawyer (1962) by direct consideration of the Bethe–Salpeter equation. As $t \to -\infty$, $K(t) \to 0$ so that at large momentum transfer the trajectory is near -1.

Questions naturally arise about the reliability of the procedure. The trajectory function (2.1.19) is not realistic. It can be shown that $K(t)$ is infinite at the normal threshold $t = 4m^2$, a behaviour which is certainly unphysical. It is natural to enquire whether power-law behaviour $s^{\alpha(t)}$ would persist if more than the leading behaviour of each diagram were to be calculated and then summed, and whether this more complete procedure would yield a more acceptable form for $\alpha(t)$. In the next section we shall develop a more powerful technique capable of answering these questions. We shall find the conclusions reassuring.

Before leaving the simple results of this section we may note an interesting physical interpretation of the series whose terms are given by (2.1.17). At fixed large s the terms first increase with n, owing to the powers of $\ln s$, but eventually decrease as $(n-1)!^{-1}$ damps them out. Stirling's formula for $n!$ at large n,

$$n! \sim (2\pi)^{1/2} n^{n+1/2} \mathrm{e}^{-n}, \qquad (2.1.20)$$

makes it easy to see that the largest terms occur for

$$n_s \sim gK(t)\ln s \qquad (2.1.21)$$

and that their value is (to within logarithmic factors) just $s^{\alpha(t)}$ with $\alpha\,(t)$ given by (2.1.19). In other words the sum of the series is completely dominated at large fixed s by just those terms given by (2.1.21) and so, in a physical interpretation, the scattering amplitude comes predominately from the ladders whose number of rungs is given by (2.1.21). Equivalently, we may say that the scattering is dominated by processes with the order of n_s intermediate particles in the s-channel.

2.2 Mellin transforms

The powerful technique we introduce in this section goes beyond the simple leading logarithmic approximations based on (2.1.7) and enables us in principle to calculate all terms in the asymptotic behaviour of a Feynman integral. It is based on the idea (Bjorken & Wu, 1963) of using Mellin transforms.

The Mellin transform $F\,(\beta)$ of a function $f(s)$ is defined by

$$F(\beta) = \int_0^\infty f(s)s^{-\beta-1}\,\mathrm{d}s. \tag{2.2.1}$$

It possesses an inversion formula

$$f(s) = \frac{1}{2\pi_\mathrm{i}} \int_C F(\beta)s^\beta\,\mathrm{d}\beta, \tag{2.2.2}$$

in which the contour C is parallel to the imaginary axis. The form of (2.2.2) makes it obvious that Mellin transforms are intimately related to the large s behaviour of $f(s)$. Clearly one should use Cauchy's theorem to move C as far as possible to the left in order to reduce the value of $\mathrm{Re}\,\beta$. The leading asymptotic behaviour will then result from the rightmost singularity in β which prevents this translation proceeding further. The argument is just the same as that employed for the Sommerfeld–Watson integral in classical Regge theory (Collins, 1977). A multiple-pole singularity in β

$$1/(\beta - \beta_0)^n \tag{2.2.3}$$

will give a term in the asymptotic behaviour

$$s^{\beta_0}(\ln s)^{n-1}/(n-1)! \tag{2.2.4}$$

The utility of Mellin transforms lies in the fact that the transform of a

Feynman integral can readily be evaluated if the integral is written in one of the exponential forms discussed in section 1.2. This appears particularly straightforward if one uses the complex exponential representation discussed after equation (1.2.33) since the latter is defined for all real values of the invariants. However some subtleties are disguised by such an approach. We prefer therefore to use the representation (1.2.29). In that case more discussion is necessary. The reader not unduly concerned about mathematical nicety can omit what immediately follows and make his way straight to equation (2.2.6).

Initially (1.2.2a) is only valid in the Euclidean region for which we know that $D \leqslant 0$. Clearly this is of no immediate use since (a) the external p_i^2 are negative there, and (b) the permitted values of s and t are negative and bounded so that $|s| \to \infty$ lies outside the region. Point (a) is dealt with by considering the possibility of continuing the p_i^2 to physical values and still retaining $D \leqslant 0$ for a bounded range of values of s and t. The region in s and t so defined is called the Symanzik region (details are given in Eden et $al.$ 1966, ch. 2) and it exists for all Feynman diagrams describing two particle scattering. If we now restrict ourselves to $planar$ diagrams we shall still have $D \leqslant 0$ if we take s out of the Symanzik region to $-\infty$, since the coefficient of s is positive definite for planar diagrams so that making s more negative can only make D more negative. This means that we may use (1.2.29) for these diagrams for the limit $s \to -\infty$, provided t is fixed at a value in the Symanzik region. We should then Mellin transform with respect to

$$\sigma = -s. \tag{2.2.5}$$

Once the asymptotic behaviour has been established by this method it can be analytically continued to limits of greater physical interest, in particular $s \to +\infty$. (Implicit in this statement is the assumption that there are no essential singularities at infinity in the s-plane which could give behaviour like $e^{\lambda s}(\lambda > 0)$, which is negligeable for negative large s and dominant for positive large s. Such behaviour is excluded by the sort of polynomial bounds for all large $|s|$ which can be constructed from the leading-order estimates of section 2.1.)

The application of Mellin transforms to non-planar diagrams will be discussed in section 2.6. We shall find that they require for their proper definition the idea of signature which plays so large a role in Regge theory (Collins, 1977).

In the planar case, therefore, we use (2.2.1) to Mellin transform (1.2.29) with respect to $\sigma = -s$. This gives

$$L(\beta,t) = (-1)^m G \int_0^\infty d\sigma\, \sigma^{-\beta-1} \int_0^\infty \prod d\tilde{\alpha}\, C(\tilde{\alpha})^{-2}$$
$$\times \exp[-g(\tilde{\alpha})\sigma/C(\tilde{\alpha})]\exp[-J(\tilde{\alpha},t)], \quad (2.2.6)$$

where $g(\tilde{\alpha})$ is the coefficient of s in D,

$$-C(\tilde{\alpha})J(\tilde{\alpha},t) = D(s=0,t,\tilde{\alpha}), \quad (2.2.7)$$

and G represents all the numerical factors including coupling constants. Interchanging the orders of integration in (2.2.6) enables the σ-integration to be performed trivially by using the identity for the gamma function,

$$\Gamma(m) = \int_0^\infty d\rho\, \rho^{m-1} e^{-\rho}.$$

We find

$$L(\beta,t) = \Gamma(-\beta)(-1)^m G \int_0^\infty \prod d\tilde{\alpha}[g(\tilde{\alpha})]^\beta [C(\tilde{\alpha})]^{-2-\beta} e^{-J(\alpha,t)}. \quad (2.2.8)$$

In inverting a Mellin transform it is necessary to determine the correct position of the contour C of (2.2.2). The poles of the gamma function in (2.2.8) at $\beta = 0,1,2,\ldots$ must lie to the right of C. This is because we know from the estimates of section 2.1 that the asymptotic behaviour of Feynman diagrams contains only negative powers of s. Thus the singularities which count for determining the asymptotic behaviour of the Feynman integral can only be those which arise from the integral in (2.2.8). J is positive definite and gives strong convergence at $\tilde{\alpha} = \infty$. The singularities of the integral therefore arise from the lower end points $\tilde{\alpha} = 0$, where divergences at negative values of β can occur from the vanishing of $g(\alpha)$. The way this can happen is illustrated by the case of the ladder diagrams of fig. 2.1.1 which we now proceed to reanalyse using these new methods.

Ladder diagrams

Application of (2.2.8) to fig. 2.1.1 yields

$$L_n(\beta,t) = \Gamma(-\beta)g^2\left(\frac{-g^2}{16\pi^2}\right)^{n-1}(-1)^n \int_0^\infty d\alpha_1\ldots d\alpha_n$$
$$\times \int_0^\infty \prod d\beta_i d\beta_i' (\alpha_1\alpha_2\ldots\alpha_n)^\beta [C(\alpha,\beta,\beta')]^{-2-\beta} e^{-J},$$
$$(2.2.9)$$

From (2.2.9) we can obtain the complete asymptotic behaviour of ladder diagrams (Polkinghorne, 1964).

The integral (2.2.9) is well behaved for real $\beta > -1$ but the α_i-integration diverges at $\alpha_i = 0$ when $\beta = -1$. The singularities corresponding to these divergences can be exhibited by integrating by parts with $\mathrm{Re}\,\beta > -1$ and then continuing in β. This replaces (2.2.9) by the equivalent expression

$$L_n(\beta,t) = \Gamma(-\beta)g^2\left(\frac{-g^2}{16\pi^2}\right)^{n-1}\int_0^\infty d\alpha_1 \ldots d\alpha_n \int_0^\infty \prod d\beta_i\, d\beta_i'$$

$$\times \frac{(\alpha_1 \ldots \alpha_n)^{\beta+1}}{(\beta+1)^n}\frac{\partial^n}{\partial\alpha_1 \ldots \partial\alpha_n}(C^{-2-\beta}e^{-j}). \qquad (2.2.10)$$

This integral is convergent for $\mathrm{Re}\,\beta > -2$ and the presence of an n-fold pole at $\beta + 1 = 0$ is explicit. The residue of this pole is readily evaluated by putting $\beta = -1$ elsewhere in the integral. The α_i-integrations can then be trivially performed with the effect of putting each $\alpha_i = 0$. We obtain

$$L_n \sim \frac{g^2\left(\frac{-g^2}{16\pi^2}\right)^{n-1}(-1)^n}{(\beta+1)^n}\int_0^\infty \prod d\beta_i\, d\beta_i'\, c^{-1}e^{\delta/c}, \qquad (2.2.11)$$

with c and δ defined by (2.1.12) and (2.1.13). The asymptotic behaviour corresponding to this leading singularity is immediately recognised to be just (2.1.14) since the integral in (2.2.11) is that associated with fig. 2.1.2 in two dimensions. However it is possible with (2.2.10) to do much better than be content with the leading singularity $(\beta + 1)^{-n}$. All the lower-order poles at $\beta = -1$ can also be evaluated to give a complete expression for the amplitude. This we now proceed to do. It is achieved by rewriting the factors in the integrand

$$\frac{\alpha_i^{\beta+1}}{\beta+1}\frac{\partial}{\partial\alpha_i} = \left(\frac{a_i^{\beta+1}-1}{\beta+1}\right)\frac{\partial}{\partial\alpha_i} + \frac{1}{\beta+1}\frac{\partial}{\partial\alpha_i}. \qquad (2.2.12)$$

The first factor on the right hand side of (2.2.12) is not singular at $\beta = -1$. The second is singular but the corresponding α_i – integration can trivially be performed with the effect of putting $\alpha_i = 0$. This just corresponds to contracting a rung of the ladder to produce a contracted diagram consisting of two subdiagrams hinged at a single vertex. In consequence the functions e^{-j} and C factorise into two parts corresponding to these two subdiagrams. Thus all $(\beta + 1)^{-1}$ singularities are associated with contractions of this sort and with consequential factorisations. If these ideas are applied to (2.2.10) and a sum then taken over n one finds that the complete set of ladder diagrams contributes

$$L(\beta,t) = \sum_{m=0}^{\infty} \Gamma(-\beta)G(\beta,t)[F(\beta,t)]^m G(\beta,t)(\beta+1)^{-m-1}$$

$$= G(\beta,t)\{\Gamma(-\beta)/[\beta+1-F(\beta,t)]\}G(\beta,t), \qquad (2.2.13)$$

where

$$F(\beta,t) = \sum_{k=0}^{\infty} -\left(\frac{-g^2}{16\pi^2}\right)^k \int_0^{\infty} d\alpha_1 \ldots d\alpha_k \int_0^{\infty} \prod d\beta_i d\beta_i'$$

$$\times \prod_{i=1}^{k}\left(\frac{\alpha_i^{\beta+1}-1}{\beta+1}\right)\frac{\partial^k}{\partial\alpha_1 \ldots \partial\alpha_k}[C_k^{-2-\beta}e^{-J_k}], \qquad (2.2.14)$$

with C_k and J_k the Feynman functions for the contracted k-runged interior subdiagram of fig. 2.2.1, and $G(\beta,t)$ is similarly defined in terms of the k-runged end subdiagrams of fig. 2.2.2. Clearly (2.2.13) corresponds to

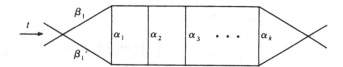

FIGURE 2.2.1 An interior subdiagram.

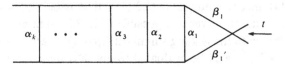

FIGURE 2.2.2 An end subdiagram.

Regge pole behaviour $s^{\alpha(t)}$, where $\alpha(t)$ is a root of the transcendental equation,

$$\alpha(t) + 1 - F(\alpha(t),t) = 0, \qquad (2.2.15)$$

and the Gs are the functions that couple the Regge pole to the external particles. When β takes non-negative integral values the poles of the Γ-function in the numerator of (2.2.13) will just correspond to the bound state poles associated with the Regge trajectory.

Equation (2.2.15) can be thought of as an alternative way of formulating the Bethe–Salpeter equation in the ladder approximation, a point of view which has been emphasised by Swift & Tucker (1970). It is convenient sometimes to rewrite (2.2.14) by integrating by parts with respect to the

α_i to give

$$F(\beta,t) = \frac{\bar{F}(\beta,t)}{1 + \bar{F}(\beta,t)/(\beta + 1)}, \qquad (2.2.16)$$

where

$$\bar{F}(\beta,t) = \sum_{k=0}^{\infty} \left(\frac{g^2}{16\pi^2}\right)^k \int_0^\infty d\alpha_1 \ldots d\alpha_k \int_0^\infty \prod d\beta_i d\beta_i' \prod_{i=1}^{k} \alpha_i^\beta C_k^{-2-\beta} e^{-J_k}. \qquad (2.2.17)$$

Solving (2.2.15) is then equivalent to finding the poles $\beta = \alpha(t)$ of $\bar{F}(\beta,t)$.

2.3 End point contributions

In the preceding sections we saw that the significant asymptotic behaviour of a Feynman integral arises from the neighbourhood of points in the region of integration where $g(\alpha)$, the coefficient of the asymptotic variable, vanishes and which cannot be avoided by use of Cauchy's theorem distortion. In the particular case of ladder diagrams the vanishing of g comes about by setting $\alpha_i = 0$, that is, it corresponds to an end point of the region of integration. We now go on to consider more elaborate examples in which g only vanishes when several αs are simultaneously set equal to zero. These are, then, further illustrations of end point contributions to asymptotic behaviour. The contributions associated with g vanishing in the interior of the region of integration (which we have seen in section 1.2 can only arise for non-planar diagrams) will be discussed in section 2.6.

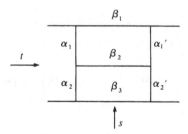

FIGURE 2.3.1 An H-diagram.

As our first example we consider the diagram of fig. 2.3.1. The coefficient of s is

$$\alpha_1 \alpha_1'(\alpha_2 + \alpha_2' + \beta_2 + \beta_3) + \alpha_2 \alpha_2'(\alpha_1 + \alpha_1' + \beta_1 + \beta_2) + \alpha_1 \beta_2 \alpha_2' + \alpha_2 \beta_2 \alpha_1'. \qquad (2.3.1)$$

To make this vanish at least two Feynman parameters must be set equal to zero; either α_1 and α_2, or α'_1 and α'_2. Clearly this corresponds to the fact that it is necessary to short-circuit two lines (either one side of the H or the other) to produce a contracted diagram which is independent of s. In these circumstances a convenient technical trick is to enforce the simultaneous vanishing of the two αs by using a change of variables called a *scaling transformation*:

$$\left.\begin{array}{l} \alpha_1 = \rho\bar{\alpha}_1, \quad \alpha_2 = \rho\bar{\alpha}_2, \quad \bar{\alpha}_1 + \bar{\alpha}_2 = 1, \\ d\alpha_1\,d\alpha_2 \rightarrow \rho\,d\rho\,d\bar{\alpha}_1\,d\bar{\alpha}_2\delta(\bar{\alpha}_1 + \bar{\alpha}_2 - 1). \end{array}\right\} \tag{2.3.2}$$

The advantage of this is that the simultaneous vanishing α_1 and α_2 is enforced by putting a single parameter, ρ, equal to zero. Such scaling transformations (of which (1.2.30) was another example) play a useful role in many manipulations. Of course a similar scaling variable ρ' is introduced for α'_1 and α'_2. Then g can be written in the form

$$g = \rho\rho'\bar{g}. \tag{2.3.3}$$

If we are only concerned with the leading logarithmic behaviour it is permissible to put $\rho = \rho' = 0$ within the rest of the integrand including $\bar{g}$, which then becomes

$$\bar{g} = \bar{\alpha}_1\bar{\alpha}'_1(\beta_2 + \beta_3) + \bar{\alpha}_2\bar{\alpha}'_2(\beta_1 + \beta_2) + \bar{\alpha}_1\beta_2\bar{\alpha}'_2 + \bar{\alpha}_2\beta_2\bar{\alpha}'_1. \tag{2.3.4}$$

This construction of $\bar{g}$ is called *linearisation* since it removes from g terms of higher order than the first in ρ and ρ'. The high energy behaviour associated with fig. 2.3.1 can now be extracted with the aid of (2.1.7) applied to the ρ- and ρ'-integrations. We find

$$K'(t)s^{-2}\ln s, \tag{2.3.5}$$

where $K'(t)$ is proportional to

$$\int_0^1 \prod d\beta\,\delta(\textstyle\sum\beta - 1) \int_0^1 d\bar{\alpha}_1\,d\bar{\alpha}_2\,\delta(\bar{\alpha}_1 + \bar{\alpha}_2 - 1) \int_0^1 d\bar{\alpha}'_1\,d\bar{\alpha}'_2\,\delta(\bar{\alpha}'_1 + \bar{\alpha}'_2 - 1)$$

$$\times\, c(\beta)/(\bar{g})^2 d(\beta,t). \tag{2.3.6}$$

Here c and d are the Feynman functions of the contracted diagram fig.

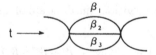

FIGURE 2.3.2 The contracted diagram associated with fig. 2.3.1.

2.3.2 and $\bar{g}$ is given by (2.3.4). In obtaining this *leading* asymptotic behaviour it is only the *minimum* sets of αs making $g = 0$ which are significant.
For example $g = 0$ if $\alpha_1 = \beta_2 = \alpha_2'$ but this has no effect in the leading
behaviour, as is shown by the fact that the integral (2.3.6) is convergent
in the neighbourhood of $\bar{\alpha}_1 = \beta_2 = \bar{\alpha}_2' = 0$.

d-lines

The general formula (2.1.7) enables us readily to extend this procedure
to the case where it is necessary to short-circuit m lines before the diagram
contracts to one independent of s. Such sets of lines are called *d*-lines of
length m (Halliday, 1963; Tiktopoulos, 1963). Assuming m to be the minimum length of *d*-lines possible for the diagram, and that there are n
independent such paths, one would expect the leading asymptotic behaviour to be proportional to

$$(1/s^m)\,(\ln s)^{n-1} \tag{2.3.7}$$

where the coefficient multiplying (2.3.7) is an integral analogous to (2.3.6)
and is obtained by using scaling transformations, linearisation, and (2.1.7).
This conclusion is, however, subject to possible modification for a number
of reasons:

(a) A rather trivial modification occurs if m is equal to p, the power of
the denominator in the original Feynman integral. In that case the natural
behaviour of the integral is already $s^{-p} = s^{-m}$, and the effect of the vanishing of g is to enhance this by one more power of the logarithm than
indicated by (2.3.7), so that the asymptotic behaviour becomes proportional to

$$(1/s^m)(\ln s)^n \tag{2.3.8}$$

This corresponds to the modification of (2.1.7) when $p = m$, noted in
section 2.1. If $p < m$ then the asymptotic behaviour is dominated by the
natural behaviour s^{-p} and the vanishing of g has no extra effect at all.

(b) The integral multiplying (2.3.7) may prove to be divergent. This
then indicates that the true asymptotic behaviour has not been extracted.
One finds that if the integral contains logarithmic divergences then
extra powers of $\ln s$ appear in the true asymptotic behaviour. If the integral
is power-law divergent then (2.3.7) is found to estimate the power of s
incorrectly. The principal source of these divergences is what are called
singular configurations. They form a topic of sufficient importance to
justify a section of their own, which follows this one. There is also a

second source of divergence which can be discussed forthwith. It arises from the fact that scalings which are not present in the original g can become possible due to the effect of linearisation. In that case they will not correspond to d-lines associated with contractions of the Feynman diagram but to certain *disconnected scaling sets* (Greenman, 1965) whose significance is not immediately recognisable from an inspection of the diagram. An example of this is given in the next section but the mathematical point can be made by considering the model integral

$$\int_0^1 \frac{dx\, dy}{[(xy + x^2)s + d]^2}. \tag{2.3.9}$$

The coefficient of s vanishes when $x = 0$ and one might suppose (2.3.9) had the asymptotic behaviour s^{-1} in consequence. However, because the x^2 term is neglected according to the idea of linearisation, a further possibility $y = 0$ also exists and the true asymptotic behaviour is proportional to $s^{-1} \ln s$. Special care has to be taken in calculating the numerical coefficient multiplying the asymptotic form in cases when extra scalings arise from linearisation; for example a naive evaluation of (2.3.9) suggests d^{-1} for this coefficient but explicit integration shows that it is $\frac{1}{2}d^{-1}$. General rules for obtaining the correct numerical coefficient have been given by Hamprecht (1965).

Independent scaling sets

Another factor that determines the numerical coefficient multiplying the leading asymptotic form is the need to enumerate correctly independent scaling sets where they occur. The point can be illustrated by considering a g-function of the form

$$g = \alpha_1 \alpha_2 + \beta_1 \beta_2. \tag{2.3.10}$$

First one can scale

$$\rho_1 : \alpha_1, \beta_1, \tag{2.3.11}$$

with the associated δ-function

$$\delta(\bar{\alpha}_1 + \bar{\beta}_1 - 1). \tag{2.3.12}$$

If we take for the second scaling

$$\rho_2 : \bar{\alpha}_1, \bar{\beta}_2, \tag{2.3.13}$$

then linearisation in (2.3.12) enforces.

$$\bar{\beta}_1 = 1, \tag{2.3.14}$$

and the sequence can be completed with the third scaling

$$\rho_3 : \bar{\beta}_2, \alpha_2. \qquad (2.3.15)$$

However at the second scaling stage one could instead have chosen

$$\rho_2' : \bar{\beta}_1, \alpha_2, \qquad (2.3.16)$$

enforcing

$$\bar{\alpha}_1 = 1, \qquad (2.3.17)$$

and completed by

$$\rho_3' : \bar{\alpha}_2, \beta_2. \qquad (2.3.18)$$

It is clear from (2.3.14) compared with (2.3.17) that these two scaling sequences are independent. They will contribute additively to the asymptotic behaviour. Again general rules have been given by Hamprecht (1965).

Generalised ladders

Finally we note how the results of this section enrich the structure of Regge poles beyond that given by the simple ladder diagrams.

Regge poles can be associated with generalised ladders, that is, infinite iterations of two-particle irreducible subdiagrams. 'Irreducible' means that they can not be subdivided into two sections joined only by two propagators. The complete theory requires the iteration of all such possible irreducible subdiagrams in all possible orderings. Some instructive points can be noted by considering particular examples.

Consider, for example, the iteration of the H-diagrams of fig. 2.3.1. In the leading logarithm approximation the behaviour of a diagram formed of n Hs is readily found to be

$$[K'(t)]^n [K(t)]^{n-1} s^{-2} (\ln s)^{2n-1} / (2n - 1)!, \qquad (2.3.19)$$

with $K'(t)$ and $K(t)$ defined by (2.3.6) and (2.1.18). The presence of only odd powers of $\ln s$ means that summation of (2.3.19) over n yields a pair of Regge poles:

$$\tfrac{1}{2} [K'(t)/K(t)]^{1/2} [s^{\alpha_+(t)} + s^{\alpha_-(t)}], \qquad (2.3.20)$$

where

$$\alpha_\pm(t) = -2 \pm [K'(t)K(t)]^{1/2}. \qquad (2.3.21)$$

Notice that $\alpha_{\pm}$ are both in the neighbourhood of -2 as $t \to -\infty$ and that this is because the iterated Hs contain d-lines of length 2.

A second case of interest is furnished by the iteration of single rung exchanges and Hs in all possible combinations, as illustrated in fig. 2.3.3.

FIGURE 2.3.3 Iterated rungs and Hs.

The rungs provide d-lines of length 1 and so control the dominant high energy behaviour. The Hs modify the form of the Regge trajectory so that the dominant behaviour in the leading logarithmic approximation corresponds to $s^{\alpha(t)}$ where

$$a(t) = -1 + K(t) + \sum K_k(t), \qquad (2.3.22)$$

and $K_k(t)$ is associated with a contraction containing k Hs, as in fig. 2.3.4.

Finally we may note that it is known (Polkinghorne, 1963b) that in the case of non-planar irreducible parts like fig. 2.3.5, the corresponding

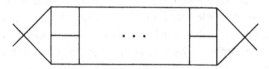

FIGURE 2.3.4 A contracted diagram giving a contribution to the leading Regge trajectory function.

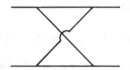

FIGURE 2.3.5 A non-planar irreducible part.

contribution to the trajectory $\alpha(t)$ possesses singularities in t additional to the expected normal thresholds. These singularities are associated with asymptotes to Landau curves (Islam, Landshoff & Taylor, 1963).

2.4 Singular configurations

The d-lines discussed in the preceding section correspond to scaling over the parameters of open arcs of lines in a Feynman diagram such that when the lines of the arc are short-circuited the resulting contracted diagram is independent of the asymptotic variable s. Tiktopoulos (1963) pointed out that to find the true asymptotic behaviour it may also be necessary to scale over sets of parameters which include those associated with some closed loops as well as the arcs of d-line type. The need for considering these *singular configurations* arises as follows.

In the spinless theory that we are currently considering the powers of the C- and D-function in a Feynman integral differ by 2. If all the αs round a loop vanish like ρ (the appropriate scaling variable) then C and D both vanish like ρ also and this produces a net factor of ρ^{-2}. (If we use the exponential form this ρ^{-2} arises directly from the C^{-2} in (1.2.29).) The presence of the ρ^{-2} has the effect of shortening any scaling that involves all the lines in the loop by 2. For example, consider fig. 2.4.1, regarded as an

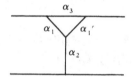

FIGURE 2.4.1 A singular configuration.

insertion in a larger diagram. It clearly has two d-line scalings of length 2 given by $\alpha_1,\alpha_2;\alpha_1',\alpha_2$. However, scaling over all the parameters $\alpha_1,\alpha_1',\alpha_2,\alpha_3$ is equally important in contributing to the asymptotic behaviour. According to the ideas already developed this extra scaling will produce an extra power of $\ln s$ in the asymptotic behaviour.

In general we can say that if the addition of $2l$ lines to a d-line arc produces l closed loops then that singular configuration scaling is as important as the original of line scaling and therefore produces a further power of $\ln s$. Fig. 2.4.1 and fig. 2.4.2a are examples of this type. If adding less than $2l$ lines produces l closed loops then the singular configuration scaling dominates, for it will give a lower power of s than the simple d-line. An example is provided by fig. 2.4.2b; the d-lines are of length 4 but the singular configuration scaling over the whole insertion is of length 3.

It is instructive to consider fig. 2.4.2a in more detail. It contributes a

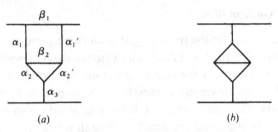

FIGURE 2.4.2 Further singular configurations.

factor to the coefficient of s given by

$$
\begin{aligned}
\alpha_1 \alpha_1'(\alpha_2 &+ \alpha_2' + \beta_2) + \alpha_1 \beta_2 \alpha_2' + \alpha_2 \beta_2 \alpha_1' \\
&+ \alpha_2 \alpha_2'(\alpha_1 + \alpha_1' + \beta_1 + \beta_2) \\
&+ \alpha_3[(\alpha_1 + \alpha_1' + \beta_1 + \beta_2)(\alpha_2 + \alpha_2' + \beta_2) - \beta_2^2].
\end{aligned} \tag{2.4.1}
$$

There are two d-lines of length 3 and two loops forming singular configurations with them. One might expect that one complete set of scalings would be

$$
\left.
\begin{aligned}
\rho_1 &: \alpha_1, \alpha_1', \alpha_2, \alpha_2', \alpha_3, \beta_1, \beta_2; \\
\rho_2 &: \alpha_1, \alpha_2, \alpha_2', \alpha_3, \beta_2; \\
\rho_3 &: \alpha_1, \alpha_2, \alpha_3; \\
\rho_4 &: \alpha_1', \alpha_2', \alpha_3.
\end{aligned}
\right\} \tag{2.4.2}
$$

However after this sequence has been performed, the residual linearised $\bar{g}$ takes the form

$$
\bar{g} = \bar{\alpha}_1 \bar{\beta}_2 \bar{\alpha}_1' + \bar{\alpha}_2(\bar{\beta}_2 \bar{\alpha}_1' + \bar{\beta}_1 \bar{\alpha}_2') + \bar{\alpha}_3 \bar{\beta}_1 \bar{\beta}_2, \tag{2.4.3}
$$

where the scaled parameters are constrained by the δ-functions

$$
\delta(\bar{\beta}_1 - 1)\delta(\bar{\beta}_2 - 1)\delta(\bar{\alpha}_1 + \bar{\alpha}_2 - 1)\delta(\bar{\alpha}_1' + \bar{\alpha}_2' + \bar{\alpha}_3 - 1). \tag{2.4.4}
$$

Equation (2.4.3) permits the further *disconnected scaling* (see section 2.3)

$$
\rho_5 : \bar{\alpha}_2, \bar{\alpha}_1', \bar{\alpha}_3 \tag{2.4.5}
$$

(Menke, 1964). Notice also that fig. 2.4.2a has two *independent scaling sets* since ρ_2 in (2.4.2) can be replaced by

$$
\rho_2' : \alpha_1', \alpha_2, \alpha_2', \alpha_3, \beta_2; \tag{2.4.6}
$$

with consequential changes in the scalings which follow.

Truss-bridge diagrams

Theories involving ϕ^4 interactions or particles with spin have a richer singular configuration structure than simple ϕ^3 theory. We shall discuss the effects of spin in section 2.8. As far as ϕ^4 theory is concerned the main effects can be illustrated by the 'truss-bridge' diagram of fig. 2.4.3 (Bjorken & Wu, 1963; we use a method of calculation taken from Polkinghorne (1976b)). All loops participate in singular configurations so that to extract the high energy behaviour it is first necessary to scale over all the Feynman parameters, then over subsets of loops and finally over individual rungs.

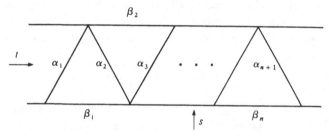

FIGURE 2.4.3 A truss-bridge diagram.

There are many independent scaling sets which contribute and they each lead to the same contribution to the asymptotic behaviour. Thus we find the leading asymptotic behaviour of the n-loop truss-bridge is

$$C_n g^2 (\bar{g})^n s^{-1} (\ln s)^{2n}/(2n)!, \qquad (2.4.7)$$

where g is the ϕ^3 coupling constant, $\bar{g}$ is the ϕ^4 coupling constant and C_n is the number of independent scaling sets in the n-loop diagram. (2.4.7) contains no dependence on t. This is because scaling over all the parameters of the diagram contracts it down to a point and removes all dependence of the coefficient upon external variables.

It is necessary to evaluate C_n. Counting the independent scaling sets involves just counting the different ways in which the loops can be scaled since every scaling set will end by scaling over the individual rungs whatever loop scalings have gone before. C_2 is 2 since we scale the whole diagram and then over one or other of the single loops. C_3 is 5. Four of these scaling sets arise from the different ways of making nested scalings over, successively, three loops, two loops and one loop. For example, labelling the loops 1, 2, 3 from left to right, one can scale over $1 + 2 + 3$, $1 + 2$, 1. The final scaling set in the three-loop diagram corresponds to scaling overall followed by two disjoint single-loop scalings, that is, to $1+2+3,1,3$, in the notation just introduced. In higher-order loop diagrams the sequence of scalings can be generated in a similar way with the nesting or disjoint alternatives entering at each level. A little thought shows that this leads to the recurrence relation

$$C_n = C_{n-1} + C_{n-2}C_1 + C_{n-3}C_2 + \ldots + C_1 C_{n-2} + C_{n-1}. \quad (2.4.8)$$

The solution of (2.4.8) is

$$C_n = (2n)!/n!(n+1)!. \quad (2.4.9)$$

Summed over n, (2.4.7) and (2.4.9) yield a behaviour proportional to $g^2 s^{-1} I_1(\bar{g}^{1/2} \ln s)/\bar{g}^{1/2} \ln s$, where I_1 is a standard Bessel function, giving an asymptotic behaviour in this leading logarithm approximation proportional to

$$g^2 s^{2\bar{g}^{1/2}-1}/\bar{g}^{3/4}(\ln s)^{3/2}. \quad (2.4.10)$$

This behaviour corresponds to a fixed-cut singularity in the Regge plane at $l = 2\bar{g}^{1/2} - 1$. The presence of fixed-cut singularities of this type is characteristic of renormalisable field theories like ϕ^4 and reflects the more singular character of the interactions they represent compared with a softer super-renormalisable field theory like ϕ^3.

2.5 Divergences of Feynman integrals

Many Feynman integrals contain divergences which must be removed before they can be given a well-defined meaning. Basic texts on quantum field theory (for example, Bjorken & Drell, 1965) give accounts of the regularisation procedures and show how they may be interpreted in renormalisable field theories as redefinitions of the masses and coupling constants. It is not the purpose of this monograph to give a detailed account of these topics but in this section we indicate how the techniques of the chapter can be used to discuss these problems.

The presence of an overall divergence in a diagram, which is due to all the loop momenta taking large values, is indicated by the pole of the Γ-function in expressions like (1.2.21) when $(n-2l)$ is non-positive. There may also be divergences in subintegrations corresponding to a subset only of the loop momenta being large. These manifest themselves as divergences at $\rho = 0$, where ρ is a scaling variable that puts to zero all the αs round the loops with large momenta. If the exponential representation (1.2.29) is used, a unified picture is obtained in which the overall

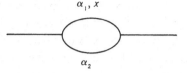

FIGURE 2.5.1 A self-energy part.

divergence is manifested by scaling over all $\tilde{\alpha}$s and the subdivergences by scaling over subsets of $\tilde{\alpha}$s.

A simple illustration is provided by fig. 2.5.1 which is the only divergent diagram in super-renormalisable ϕ^3 theory. To within numerical factors the contribution is

$$\int_0^\infty d\tilde{\alpha}_1 d\tilde{\alpha}_2 (\tilde{\alpha}_1 + \tilde{\alpha}_2)^{-2} \exp\left[(\tilde{\alpha}_1\tilde{\alpha}_2 p^2 - (\tilde{\alpha}_1 + \tilde{\alpha}_2)^2 m^2)/(\tilde{\alpha}_1 + \tilde{\alpha}_2)\right], \quad (2.5.1)$$

with momentum p running through the diagram. Scaling $\tilde{\alpha}_1, \tilde{\alpha}_2$ with ρ gives a logarithmic divergence

$$\int_0^\infty \frac{d\rho}{\rho}, \quad (2.5.2)$$

corresponding to the logarithmically divergent loop integral.

A method of great elegance and utility in the discussion of divergences is provided by *dimensional regularisation* (Bollini & Giambiagi, 1972; 't Hooft & Veltman, 1972). The idea is to continue analytically in d, the dimensionality of space-time. Divergences then manifest themselves as poles at the physical value $d = 4$.

Changing the dimension of space to d in (1.2.29) implies the replacement

$$C(\tilde{\alpha})^{-2} \to C(\tilde{\alpha})^{-d/2}. \quad (2.5.3)$$

Introducing the scaling variable ρ, (2.5.1) now takes the form

$$\int_0^\infty d\rho \rho^{-d/2+1} \int_0^1 d\alpha_1 d\alpha_2 \delta(\alpha_1 + \alpha_2 - 1) \exp\left[\rho(\alpha_1\alpha_2 p^2 - m^2)\right]. \quad (2.5.4)$$

The pole at $d = 4$ is exhibited by integrating by parts to give

$$\frac{2}{4-d} - \int_0^\infty d\rho \ln \rho \int_0^1 d\alpha_1 d\alpha_2 \delta(\alpha_1 + \alpha_2 - 1)$$

$$\times (\alpha_1\alpha_2 p^2 - m^2) \exp\left[\rho(\alpha_1\alpha_2 p^2 - m^2)\right]$$

$$+ \text{terms } O(d-4). \quad (2.5.5)$$

The prescription is then to cancel the pole by a renormalisation counter-term and to retain the finite part at $d = 4$ given by (2.5.5). In fact in the application of these ideas it is customary to introduce further d-dependence above that given by (2.5.3). For example the dimensionless coupling constants of renormalisable field theories acquire dimensions when $d \neq 4$ so that it is usual to write

$$\bar{g} \rightarrow \bar{g}(\mu)^{4-d}, \qquad (2.5.6)$$

where μ is some appropriate mass. The effect of adding such terms as (2.5.6) is readily seen to be equivalent to the addition of a constant to the finite part of the renormalised integral.

A more old-fashioned treatment of divergences is to introduce Feynman cut-offs

$$\frac{1}{q^2 - m^2} \rightarrow \frac{1}{q^2 - m^2} \frac{-\Lambda^2}{q^2 - \Lambda^2}, \qquad (2.5.7)$$

associated with sufficient lines in the diagram to make all integrations convergent. Λ parametrises the cut-off and the behaviour as $\Lambda^2 \rightarrow \infty$ exhibits the divergence of the diagram. This behaviour is readily calculated by using our techniques with Λ^2 being the asymptotic variable. Extra Feynman parameters x_i must be introduced corresponding to the extra denominators introduced by (2.5.7). The leading Λ^2 behaviour will be obtained by scaling round the divergent loops (including the parameters x_i associated with them) and also setting the $x_i = 0$. In fig. 2.5.1, for example, it is only necessary to make the substitution (2.5.7) for one of the lines, say the one associated with α_1. The scalings α_1, α_2, x; then x; give an asymptotic behaviour $\ln \Lambda^2$ corresponding to the logarithmic divergence of the diagram.

2.6 Pinch contributions

So far we have only considered contributions to the asymptotic behaviour that arise from the vanishing of the coefficient of the asymptotic variable, $g(\alpha)$, at an edge of the region of integration. It is also possible to have contributions that arise from the vanishing of g in the interior. If this vanishing is to be effective it is necessary that the points at which it occurs cannot be avoided by a simple displacement of the integration contours such as is permitted by Cauchy's theorem. The condition for this is that g can be decomposed into two factors, each of which vanishes separately. To understand this condition, extract a factor of s from D so that it takes the form

$$g(\alpha) + d(\alpha, t)/s. \tag{2.6.1}$$

The asymptotic behaviour can be thought of as due to a singularity at $s = \infty$, which makes (2.6.1) vanish there when $g = 0$. If this singularity is to be inescapable, so that Cauchy's theorem will not enable the contour to be distorted to avoid it, then in fact it must be a coincidence of two singularities, with the contour trapped between them. That is, the two singularities must pinch the contour. (This idea is familiar in the study of the finite plane singularities of Feynman integrals; see Eden *et al.*, 1966, ch. 2.) If g factors into two separately vanishing terms the double root at $s^{-1} = 0$ of the equation given by the vanishing of (2.6.1) provides the possibility of just such a pinch. The resulting contributions to high energy are called *pinch contributions* (Polkinghorne, 1963*c*; Tiktopoulos, 1963). An alternative approach to them is given later in this section.

A simple model

An integral which illustrates this idea is provided by

$$I = \int_{x_1}^{x_2} dx \int_{y_1}^{y_2} dy \frac{1}{(xys + d)^{n+2}}$$

$$= -\frac{1}{(n+1)sd^{n+1}} \ln \frac{(x_1 y_1 s + d)(x_2 y_2 s + d)}{(x_1 y_2 s + d)(x_2 y_1 s + d)} + R(s), \tag{2.6.2}$$

where $R(s)$ is a rational function whose explicit form need not be written out. As $s \to \infty$ the logarithm in (2.6.2) tends to $\ln(1 + O(s^{-1}))$. If the principal branch of the logarithm is chosen then cancellations with $R(s)$ yield a net asymptotic behaviour of s^{-n-2}, which is the natural behaviour associated with the integral. However if a branch is chosen in which $\ln 1 = 2\pi mi$, $m \neq 0$, then

$$I \sim [-2\pi mi/(n+1)d^{n+1}]s^{-1}. \tag{2.6.3}$$

We shall show that (2.6.3) is just the pinch asymptotic behaviour.

If $x_1, \ldots y_2$, are all positive then the principal branch is correct since all the logarithms in (2.6.2) are evaluated with positive arguments and take their principal values. There is no contribution (2.6.3) because the pinch at $x = y = 0$ is outside the contour of integration. If $x_1, y_1, < 0$ and $x_2, y_2 > 0$ then it is easy to see that $m = -1$. (The numerator factors in the logarithm tend to $+\infty$ and the denominator factors to $-\infty$.

To evaluate the branch of the latter we assume that d contains a $+i\varepsilon$ prescription.) In this case $I \sim s^{-1}$. This is because the contour is now trapped at $x = y = 0$.

Iterated crosses

In Feynman integrals pinch contributions only arise from non-planar diagrams since in planar diagrams g is a sum of products of the αs and can only vanish when a set of αs vanishes. The very simplest example of a pinch contribution is provided by the iterated cross of fig. 2.6.1. The coefficient of s at fixed t is

$$(\alpha_1 \alpha_3 - \alpha_2 \alpha_4)(\alpha_1' \alpha_3' - \alpha_2' \alpha_4'). \tag{2.6.4}$$

End point behaviour requires two αs to vanish and gives $s^{-2} (\ln s)^5$. However, a straightforward application of the model (2.6.2), with $x = \alpha_1 \alpha_3 - \alpha_2 \alpha_4$, and $y = \alpha_1' \alpha_3' - \alpha_2' \alpha_4'$, shows that the leading asymptotic behaviour is s^{-1}, coming from the pinch contribution. An iteration of n-crosses would give a leading pinch behaviour proportional to $s^{-1} (\ln s)^{n-2}$.

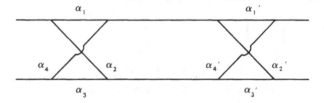

FIGURE 2.6.1 Iterated crosses.

In the Mellin transform plane the pinch contributions of the n-fold crosses correspond to a sequence of poles $(\beta + 1)^{-n+1}$. Their residues do not factorise and the sum must be interpreted as corresponding to an essential singularity at $\beta = -1$ (Contogouris, 1965; Kaschlun & Zoellner, 1965). The existence of this essential singularity had been predicted by Gribov & Pomeranchuk (1962) on the basis of a contradiction with the requirement of two-particle unitarity in the t-channel for theories with third Mandelstam spectral functions. (See the discussion in section 3.7.)

An alternative approach

It is possible to reduce the calculation of pinches to end point behaviour by decomposing the region of integration into sectors in which $g(\alpha)$

has a definite sign. This is essential if we are to use the Mellin transform techniques of section 2.2, for in the regions where $g > 0$, $-s$ is the natural asymptotic variable (because it preserves the negative definite property of D in (1.2.26)), but in regions where $g < 0$, s is the natural asymptotic variable (for a similar reason). The regions are bounded by $g = 0$ and this new boundary generates its own end point behaviour. The combination of these end point behaviours then yields the pinch contributions calculated previously.

For example, consider the model integral (2.6.2). Assuming $x_2, y_2 > 0$; $x_1, y_1 < 0$, we can divide the integration region up into the two sectors

$$\left. \begin{aligned} x_2 \geqslant x \geqslant 0, \quad y_2 \geqslant y \geqslant 0; \\ 0 \geqslant x \geqslant x_1, \quad 0 \geqslant y \geqslant y_1; \end{aligned} \right\} \tag{2.6.5}$$

in which the coefficient of s is positive; and the two sectors

$$\left. \begin{aligned} x_2 \geqslant x \geqslant 0, \quad 0 \geqslant y \geqslant y_1; \\ 0 \geqslant x \geqslant x_1, \quad y_2 \geqslant y \geqslant 0; \end{aligned} \right\} \tag{2.6.6}$$

in which the coefficient of s is negative. The two regions (2.6.5) together yield an asymptotic behaviour

$$-[2/(n+1)d^{n+1}](-s)^{-1}\ln(-s), \tag{2.6.7}$$

from the end point $x = y = 0$, and two regions (2.6.6) an asymptotic behaviour

$$-[2/(n+1)d^{n+1}](s)^{-1}\ln(s), \tag{2.6.8}$$

for the same end points. Combining (2.6.7) and (2.6.8) yields the net asymptotic behaviour $2\pi i s^{-1}/(n+1)d^{n+1}$ which is just the pinch contribution.

Another possible way of discussing pinch behaviour is to introduce signatured Mellin transforms (Polkinghorne, 1968), in analogy with signature for Regge poles (see Collins, 1977). The double cross diagram is then found to correspond to a double pole at $\beta = -1$ in the even (wrong) signatured amplitude.

2.7 Regge cuts

Regge pole singularities associated with generalised ladders represent the simplest singularities in complex angular momentum. The suggestion that the s-channel iteration of Regge poles might yield more complicated Regge cut singularities was first made by Amati, Fubini & Stanghellini (1962). We shall defer a full discussion of Regge cuts till section 3.5 since

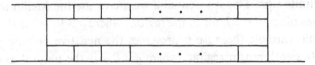

FIGURE 2.7.1 The AFS diagram.

the general theory is more readily handled by the methods of the succeeding chapter. In this section we shall note some simple results which follow from the study of Feynman integrals.

The diagram originally considered by Amati, Fubini & Stanghellini (AFS) was that shown in fig. 2.7.1. It clearly combines two Regge poles, and in section 3.5 we shall see that its discontinuity across the two-particle cut in the s-channel possesses the asymptotic behaviour.

$$\int \frac{dt_1\, dt_2\, s^{\alpha(t_1)+\alpha(t_2)-1}}{\lambda^{1/2}(t,t_1,t_2)}, \qquad (2.7.1)$$

$$\lambda(t,t_1,t_2) = t^2 + t_1^2 + t_2^2 - 2tt_1 - 2tt_2 - 2t_1 t_2, \qquad (2.7.2)$$

characteristic of a Regge cut. However this does not necessarily imply that the amplitude itself possesses this behaviour on the physical sheet since it may only manifest itself in the amplitude on the unphysical sheet reached through the two-particle normal threshold, the discontinuity being, of course, the difference of these two amplitudes. That this is in fact the case is immediately apparent on inspecting fig. 2.7.1 (Polkinghorne, 1963d). The diagram is planar so that its asymptotic behaviour on the physical sheet (that is, for positive Feynman αs) is purely end point, giving a leading behaviour

$$s^{-3}\ln s, \qquad (2.7.3)$$

however many rungs in either ladder. This is manifestly not the behaviour (2.7.1).

Clearly on the basis of such arguments only non-planar diagrams can have any chance of giving physical sheet Regge cuts. Mandelstam (1963) argued that diagrams like fig. 2.7.2a (which combines a Regge pole with a particle pole) and fig. 2.7.2b (which combines two Regge poles) would be those which would give cut contributions. This has been verified by direct calculation both for fig. 2.7.2a (Polkinghorne, 1963c) and fig. 2.7.2b (Hasslacher & Sinclair, 1971). For full details of the calculations reference can be made to the original papers or to Eden et al. (1966, section 3.8). The essential idea is to combine a pinch contribution from the crosses with end point contributions from the ladders. For example the calculation

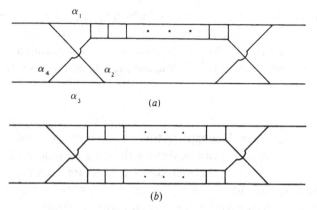

FIGURE 2.7.2 Mandelstam diagrams giving Regge cuts.

of fig. 2.7.2a can, after some algebra, be reduced to the evaluation of the leading asymptotic behaviour of an integral of the form

$$
\int_{-}^{+} dx \int_{-}^{+} dy \prod_{i=1}^{n} \int_{0}^{+} d\beta_i \frac{1}{[(xy + \prod \beta_i)s + d]^n}
$$

$$
\sim \frac{2\pi i}{(n-1)(n-2)d^{n-2}} s^{-2} \frac{(\ln s)^{n-1}}{(n-1)!},
$$

(2.7.4)

where xy is (2.6.4) formed out of the parameters of the crosses. It can then be shown that a summation of terms of this type gives the behaviour (2.7.1), with $\alpha(t_2) = 0$ for the elementary particle.

A special feature manifests itself in the calculation of the diagrams of fig. 2.7.2. The behaviour associated with (2.7.4) which builds up the Regge cut for fig. 2.7.2a corresponds to a singularity at $\beta = -2$ in the Mellin transform plane (note the s^{-2} in (2.7.4)). The crosses also possess end point scalings of length 2 which mix with the Regge cut scalings and which have also to be extracted in calculating the true asymptotic behaviour.

For example it is necessary in addition to putting to zero the β_i associated with the rungs of the ladder, to put $\alpha_3 = 0$ since this together with the condition

$$
x \equiv \alpha_1 \alpha_3 - \alpha_2 \alpha_4 = 0
$$

(2.7.5)

from the pinch reproduces the end point scaling $\alpha_3 = \alpha_2 = 0$. An analogous contribution must also be extracted from the second cross of fig. 2.7.2a. A similar phenomenon occurs for fig. 2.7.2b, where the Regge cut is built up from contributions at $\beta = -3$ which are enhanced by the d-line scalings of length 3 which the crosses possess. In both cases this minor complication can be avoided by choosing a more complicated structure for the non-planar insertions to avoid additional end point scalings.

2.8 Spin

Particles with spin introduce extra factors into the numerators of Feynman integrals. In principle this can be dealt with by the techniques described so far but in practice the algebraic complications are such that it is frequently only possible to obtain explicit results for low-order diagrams.

The extra numerator factors generate two types of effect:

(i) External momenta are present in the numerator, either initially or as a result of the displacement transformation (1.2.12). Scalar products of these momenta can give powers of the asymptotic variable. We shall return to this question below when we discuss momentum flows.

(ii) Products of the symmetric loop momenta k' present in the numerator lead to extra factors of the form (1.3.1). Taking $j_1 = j_2 = j$ in (1.3.1) gives a factor corresponding to $k_j'^2$ which behaves like ρ^{-1} when the αs round the jth loop are scaled with ρ. This enhances the possibility of singular configurations associated with the jth loop. In consequence singular configurations play a much more prominent role in theories with spin than they do for simple ϕ^3 theory.

Despite these technical difficulties it is sometimes possible to guess the likely effects due to spin. For example, consider ladder diagrams. In spinless theory the singularities of each individual ladder are at $\beta = -1$ in the Mellin transform plane, or equivalently at $l = -1$ in the Regge plane. In the weak coupling limit (that is, expanding in g^2, the coupling constant, as in (2.1.19)) the associated Regge trajectory is in the neighbourhood of $l = -1$. If the sides of the ladder are now replaced by particles of intrinsic spin s_1 and s_2 respectively, one might suppose that these angular momenta add to the 'orbital' angular momentum $l = -1$ of the ladder itself to promote the leading singularity of an individual diagram to $l = s_1 + s_2 - 1$, so that the leading log behaviour (2.1.14) would be changed to

$$s^{s_1 + s_2 - 1} (\ln s)^{n-1}/(n-1)!. \tag{2.8.1}$$

In the weak coupling limit the trajectory would now be near $l = s_1 + s_2 - 1$. This is sometimes called the translational effect of spin.

This intuition does not prove to be quite correct. When vector particles are present it is always necessary to calculate a gauge-invariant set of diagrams. It is found that within a gauge-invariant set individual diagrams can have asymptotic behaviours in excess of the estimate (2.8.1). The result expected by intuition is then only found after taking into account cancellations among diagrams of the set. In calculations it is always worth considering whether the effect of these cancellations can be minimised by a clever choice of gauge (since the asymptotic behaviour of individual diagrams is in general gauge dependent). These matters have been extensively studied because of an interesting phenomenon which we now discuss.

Reggeisation

In ϕ^3 theories the leading Regge pole in the weak coupling limit is near $l = -1$. There is also a direct t-channel pole corresponding to the single spinless particle intermediate state in that channel. This behaves like s^0 for all t and so corresponds to a fixed singularity at $l = 0$ (in fact a Kronecker-delta, δ_{0l}). There is no way in which this singularity can combine with the Regge pole near $l = -1$. Thus we have a physical picture in which the particle represented by the ϕ-field appearing in the Lagrangian corresponds to a Kronecker-delta singularity in the Regge plane and the Regge trajectories only describe particles which are bound states of two or more such elementary ϕ-particles. It is a surprising possibility, first suggested by Gell-Mann *et al.* (1964), that this separation between 'elementary' particles, corresponding to fields in the Lagrangian and not lying on Regge trajectories, and their bound states, which do lie on trajectories, does not necessarily hold in theories with vector particles. Instead, in that case it is possible for the 'elementary' particle poles to get caught up onto trajectories, that is, to become Reggeised. This is because the translational effect of spin promotes the trajectories to levels where it is possible for them to combine with the apparent Kronecker-delta singularities.

For example, in a theory with a spinor particle interacting with a vector particle, the spin $\frac{1}{2}$ pole of fig. 2.8.1a and the box diagram of fig. 2.8.1b (in which the wavy lines denote spin 1 particles) have asymptotic behaviours which differ by only a logarithm. If certain conditions were to be satisfied it would then be possible that the contribution of fig. 2.8.1b gave the first non-trivial term in the expansion of a Regge pole on whose trajectory lay the particle appearing in the intermediate state of the Born approximation diagram, fig. 2.8.1a. The corresponding trajectory function

(a)   (b)

FIGURE 2.8.1 The diagrams associated with Reggeisation.

would have to have the form

$$\alpha(t) = \tfrac{1}{2} + (t - m^2)f(t), \tag{2.8.2}$$

so that it passed through $\alpha = \tfrac{1}{2}$ at $t = m^2$. This conjecture was confirmed by calculations by Cheng & Wu (1965). Mandelstam (1965) discussed the general conditions under which Reggeisation could be expected to occur. Recently extensive calculations (Tyburski, 1976; Cheng & Lo, 1977; Grisaru 1977) have shown that Reggeisation occurs in non-Abelian gauge theories with spontaneous symmetry breaking.

Momentum flows

It is particularly important in discussing particles with spin to have as clear an understanding as possible of the way the large momentum flows through the diagram. Such an understanding also forms a link between the Feynman parameter methods of this chapter and the momentum space techniques of chapter 3. Many of the points of interest are illustrated by the discussion of large angle scattering given by Halliday, Huskins & Sachrajda (1974).

High energy wide angle scattering is a regime characterised by large s and t:

$$s \to \infty, \quad t \to \infty, \quad s/t \sim -2/(1 - \cos\theta), \quad \text{fixed}, \tag{2.8.3}$$

where θ is the centre-of-mass scattering angle. We consider the contribution in the leading logarithmic approximation from generalised ladders like fig. 2.8.2, where the sides of the ladder correspond to spin $\tfrac{1}{2}$ particles and the wavy lines represent vector gluons. (The distinction between the bold line and the rest is explained in what follows.)

It can be shown that in the leading log approximation the significant region of integration for the symmetric loop momenta is where they are all comparatively small,

$$|k''| \ll \sqrt{s}. \tag{2.8.4}$$

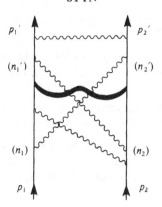

FIGURE 2.8.2 The eikonal approximation. The bold wavy line represents the hard gluon; the other wavy lines represent soft gluons.

This is because if $k''^2 \sim s$ the leading asymptotic behaviour must come from a region of Feynman αs where the coefficient of k''^2 is zero. Referring to the discussion of section 1.2 one sees that this necessarily implies that C must vanish due to scaling round the loops which contain k''. That requirement then excludes d-line scalings which are needed to obtain the maximum power of $\ln s$, so that (2.8.4) must hold to avoid this exclusion. (For an alternative insight into this condition see the discussion of logarithmic effects in section 3.3.) A consequence of (2.8.4) is that the large momenta are given by the quantities Y_i introduced in section 1.2. The rules for the evaluation of the Y_i then enable us to determine the large momentum flow associated with a sequence of scalings.

Consider for example the beginning of a sequence of scalings discussed by Halliday *et al.* (1974) which is pictured in fig. 2.8.3. Here the bold lines indicated the lines whose αs are set to zero. For the first scaling the 3s and 4s label lines whose Y_i are forced to be parallel to p_3 or p_4 respectively. Similarly the 1s and 2s in the second scaling give lines whose Y_i are parallel to p_1 and p_2. For the leftmost rung of the ladder the clash between the 4 of the first scaling and the 1 of the second implies that after both scalings are performed no large momentum can flow through the line at all.

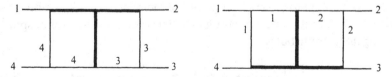

FIGURE 2.8.3 The first two steps in a scaling sequence.

In a similar way no large momentum can flow through the rightmost rung of the ladder either. Therefore any scaling containing these two steps must give the momentum flow of fig. 2.8.4a. Analysis shows that the dominant scalings always give momentum flows in which the external momenta never split but flow wholly down one line. Another pattern fulfilling this condition is shown in fig. 2.8.4b. However for spin $\frac{1}{2}$-vector gluon scattering this latter is less effective. This is because one of the spinor numerators down the side of the ladder lacks a large momentum. These large numerator factors combine to give powers of s (as we shall see) so that the absence of this factor depresses the asymptotic behaviour given by fig. 2.8.4b.

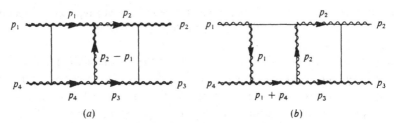

FIGURE 2.8.4 Some possible momentum flows.

We have therefore reached the important conclusion that scalings which give the flow pattern of fig. 2.8.4a will dominate in the leading log approximation for wide angle scattering. This pattern is characterised by the large momentum transfer from one side of the ladder to the other being carried by a single gluon. This is the so-called hard gluon distinguished by the bold line in fig. 2.8.2. It would be possible to calculate the contribution due to this momentum flow by using α-space scalings but instead we shall introduce another method of calculation which is of interest.

The relativistic eikonal approximation

This approximation (Lévy & Sucher, 1969; Abarbanel & Itzykson, 1969; Cardy, 1970; Tiktopoulos & Treiman, 1971) sums up the effect of a single hard vector gluon exchange together with the exchange of many soft gluons (that is, gluons carrying small momentum).

Suppose there are n_1 such soft gluons of momenta $k_1 \ldots k_{n_1}$ emitted by particle 1 before it emits the hard gluon. The corresponding set of spin $\frac{1}{2}$ propagators contributes

$$\frac{\gamma \cdot p_1 + m}{-2p_1 \cdot (k_1 + \ldots + k_{n_1})} \gamma_{\mu n_1} \ldots \gamma_{\mu 1} \frac{\gamma \cdot p_1 + m}{-2p_1 \cdot k_1} \gamma_{\mu 1} u(p_1), \qquad (2.8.5)$$

where we have put $p_1^2 = m^2$. We have also used the fact that the k_i are small so we may neglect them in the numerators and neglect k_i^2 in the denominators. Simple manipulation which makes use of the Dirac equation and the anticommutation of the γs reduces (2.8.5) to the form

$$\frac{2p_{1\mu n_1}}{-2p_1\cdot(k_1 + \ldots + k_{n_1})} \ldots \frac{2p_{1\mu_1}}{-2p_1\cdot k_1} u(p_1). \tag{2.8.6}$$

Similar manipulations can be performed on the particle 1 propagators following the hard gluon (with p_1 replaced by p_1') and on the two sets of propagators associated with particle 2. If we now permute the order in which gluon lines are attached to the particle lines and sum over resulting contributions, (2.8.6) reduces to

$$\frac{2p_{1\mu n_1} \ldots 2p_{1\mu_1}}{(-2p_1\cdot k_n) \ldots (-2p_1\cdot k_1)} u(p_1). \tag{2.8.7}$$

However, by performing such permutations on the four sets of gluon lines separately we have overcounted the same Feynman diagram many times and we must divide by

$$(n_{12})!(n_{12'})!(n_{1'2})!(n_{1'2'})!, \tag{2.8.8}$$

where n_{ij} is the number of gluon lines joining p_i to p_j. The hard gluon line carries a momentum

$$q - \sum_{i=1}^{n_1} k_i - \sum_{i=1}^{n_1'} k_i', \tag{2.8.9}$$

where

$$q^2 = t. \tag{2.8.10}$$

If we neglect the ks in (2.8.9) the k-integrations will not converge but this can be dealt with by the substitution

$$1/-2p\cdot k \rightarrow 1/(\varepsilon k^2 - 2p\cdot k), \tag{2.8.11}$$

in (2.8.7), with ε small, for we shall find the final asymptotic behaviour is independent of ε. With this substitution one can neglect the ks in (2.8.9). Thus, finally, summing over n_{ij}, one obtains a behaviour of the form

$$B(s,\theta)e^{\Sigma U_{ij}}, \tag{2.8.12}$$

where B is the Born term

$$(1/t)\bar{u}(p_1')\gamma_\mu u(p_1)\cdot \bar{u}(p_2')\gamma^\mu u(p_2), \tag{2.8.13}$$

and

$$U_{ij} = 4p_i p_j \frac{ig^2}{(2\pi)^4} \int d^4 k \frac{1}{k^2 - \mu^2} \frac{1}{\varepsilon k^2 - 2p_i \cdot k} \frac{1}{\varepsilon k^2 - 2p_j \cdot k},$$
$$i = 1, 1', \quad j = 2, 2'. \qquad (2.8.14)$$

U_{12} and $U_{1'2'}$ are functions of s; $U_{12'}$, $U_{1'2}$ are functions of u. Their asymptotic behaviours at large s and u are calculable by standard methods and are found to be independent of ε. Cancellations occur between the two sorts of Us and finally one finds a net asymptotic behaviour for fig. 2.8.2 proportional to

$$s^{(g^2/4\pi^2)\log[(1 + \cos \theta)/2]}, \qquad (2.8.15)$$

where θ is the fixed angle of scattering. The exponent is negative so that the soft gluon corrections depress the asymptotic behaviour below the behaviour given by the Born approximation term. (The latter corresponds to a constant behaviour since the spinor factors in (2.8.13) produce a factor of s.) The exponentiation of soft gluon corrections represented by (2.8.15) is very characteristic of the eikonal approximation.

However, Fig. 2.8.2 should have added to it the effect of further soft gluons which are emitted and absorbed along the sides of the ladder. They produce an additional U-function in the exponential, whose argument is t. The uncancelled $\log^2$ asymptotic behaviour of this term modifies (2.8.15), depressing the leading behaviour still further so that it becomes

$$s^{-B \ln s}, \qquad (2.8.16)$$

where B is a constant independent of θ (Cardy, 1971).

2.9 Multi-particle processes

Processes involving the production of particles possess a more complicated Regge structure than that associated with the simple two-particle scatterings considered so far. For example it is possible to exchange more than one Regge pole. The general theory is most readily developed by using the techniques of the following chapter (see section 3.6) but we shall briefly indicate some simple results which can be obtained from Feynman integrals.

Consider the $2 \to 3$ particle process

$$p_1 + p_2 \to p'_1 + p'_2 + p'_3. \qquad (2.9.1)$$

A limit of particular interest in the physical region is that in which

$$\left.\begin{aligned} s &\equiv (p_1 + p_2)^2 \to \infty, \\ s_1 &\equiv (p_1' + p_3')^2 \to \infty, \\ s_2 &\equiv (p_2' + p_3')^2 \to \infty, \end{aligned}\right\} \qquad (2.9.2)$$

with

$$\left.\begin{aligned} \eta &\equiv s/s_1 s_2, \\ t_1 &\equiv (p_1 - p_1')^2, \\ t_2 &\equiv (p_2 - p_2')^2, \quad \text{fixed.} \end{aligned}\right\} \qquad (2.9.3)$$

The fact that the two subenergies s_1 and s_2 are large and the associated momentum transfers t_1 and t_2 are fixed suggests that double Regge pole exchange may be significant and so we consider the diagram of fig. 2.9.1 with its double ladder exchange (Polkinghorne, 1965). The contribution of the large variables in D can be written in the form

$$\prod \alpha_i \prod \alpha_j' \eta s_1 s_2 + \prod \alpha_i \Delta_1 s_1 + \prod \alpha_j' \Delta_2 s_2, \qquad (2.9.4)$$

by using (2.9.3). Δ_1 and Δ_2 are functions whose explicit form need not be written out. (Their essential property is that Δ_1 does not vanish when $\alpha_j' = 0$ and Δ_2 does not vanish when $\alpha_i = 0$.) It is then possible to perform the double Mellin transform with respect to $-s_1$ and $-s_2$ by methods similar to those of section 2.2. If β_1 and β_2 are the corresponding transform variables the result takes the form of an integral over all the Feynman parameters,

$$\prod \int d\alpha_i \, d\alpha_j' \, d\delta_k \prod \alpha_i^{\beta_1} \prod \alpha_j'^{\beta_2} \, \Gamma(\beta_1, \beta_2, \eta) e^{-J/C}, \qquad (2.9.5)$$

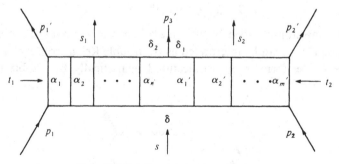

FIGURE 2.9.1 Double ladder exchange.

where

$$\Gamma(\beta_1, \beta_2, \eta) \equiv \int_0^\infty dx\, dy\, x^{-\beta_1 - 1} y^{-\beta_2 - 1}$$

$$\times \exp\{-[xy(-\eta) + x\Delta_1 + y\Delta_2]/C\}, \qquad (2.9.6)$$

$$-J = D|_{s_1 = s_2 = 0}, \qquad (2.9.7)$$

and the variables x and y are given by

$$x = (-s_1)\prod\alpha_i, \quad y = (-s_2)\prod\alpha_j'. \qquad (2.9.8)$$

The poles at $\beta_1 = -1$ due to $\alpha_i = 0$, and at $\beta_2 = -1$ due to $\alpha_j' = 0$ may be expected to generate the two Regge poles. To see how this happens we note that when all the α_i, α_j' vanish, C, Δ_1 and Δ_2 can be written as

$$\left.\begin{array}{l} C = C_0 C_0'(\delta_1 + \delta_2 + \delta), \\ \Delta_1 = C_0'\delta_1, \\ \Delta_2 = C_0\delta_2, \end{array}\right\} \qquad (2.9.9)$$

where C_0 and C_0' are the C-functions associated with the contractions of the two ladders respectively (cf. fig. 2.1.2). This enables (2.9.6) to be rewritten when the αs and α's vanish in the form

$$\Gamma_0(\beta_1, \beta_2, \eta) = C_0^{-\beta_1} C_0'^{-\beta_2} \int_0^\infty d\tilde{x}\, d\tilde{y}\, \tilde{x}^{-\beta_1 - 1} \tilde{y}^{-\beta_2 - 1}$$

$$\times \exp\{-[\tilde{x}\tilde{y}(-\eta) + \delta_1\tilde{x} + \delta_2\tilde{y}]/(\delta_1 + \delta_2 + \delta)\}, \qquad (2.9.10)$$

where

$$\tilde{x} = C_0^{-1}x, \quad \tilde{y} = C_0'^{-1}y. \qquad (2.9.11)$$

This gives a leading asymptotic behaviour of the form

$$g^2 B(t_1, t_2; \eta)(-s_1)^{\alpha(t_1)}(-s_2)^{\alpha(t_2)}, \qquad (2.9.12)$$

with the trajectories α given by the same ladder expressions as in section 2.1. Equation (2.9.7) can be interpreted as the double Regge pole exchange of fig. 2.9.2. The form is one conjectured heuristically by Kibble (1963)

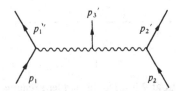

FIGURE 2.9.2 Double Regge pole exchange.

and Ter-Martirosyan (1963). B gives the coupling of the two Regge poles to the particle p'_3. It is important to note that it is a function of η as well as t_1 and t_2.

2.10 Deep inelastic scattering

The paradigm deep inelastic process is provided by electroproduction in which a photon of large (negative) mass squared q^2 scatters off a target of momentum p, in the limit of large

$$\nu \equiv p{\cdot}q. \tag{2.10.1}$$

The total cross-section for this process is given via the optical theorem by the imaginary part of the forward scattering amplitude of fig. 2.10.1.

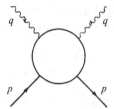

FIGURE 2.10.1 The deep inelastic scattering process.

We are concerned with the Bjorken limit (1969) in which

$$\nu, q^2 \to \infty, 2\nu/-q^2 = \omega, \quad \text{fixed.} \tag{2.10.2}$$

Of course, in this limit $s = q^2 + 2\nu + p^2$ is also large. The general discussion of behaviour in this limit is best pursued by the methods of chapter 3 and we defer a full discussion until then. In this section we simply note certain important points which arise from a study of this limit in perturbation theory. (Altarelli & Rubinstein, 1969; Chang & Fishbane, 1970; Gaisser & Polkinghorne, 1971; Mason, 1973).

Since s and q^2 are both large we may expect that so-called 'handbag' diagrams of the type of fig. 2.10.2, in which a single line joins the two current vertices, will be of particular importance. This is because putting the Feynman parameter α of this line to zero removes the dependence of the diagram on the two large variables ν and q^2. In fact, if we write the large variables in terms of ν and ω, the coefficient of ν in D for fig. 2.10.2 is of the form

$$2\nu\alpha[g - \omega^{-1}(C - \alpha\tilde{C})], \tag{2.10.3}$$

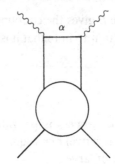

FIGURE 2.10.2 The handbag diagram.

where αg is the coefficient of s in fig. 2.10.2, C is the usual Feynman function, and $\tilde{C}$ is another function which need not concern us.

In a super-renormalisable theory like ϕ^3 the asymptotic behaviour of the amplitude in the Bjorken limit is given by the $\alpha = 0$ end point contribution of fig. 2.10.2. We find a behaviour given by

$$\frac{1}{2v}\Pi\int_0^\infty d\alpha_i \, \frac{N_{\mu\nu}(\alpha,p,q)C_0^{-2}e^{D_0/C_0}}{g/C_0 - \omega^{-1}}, \qquad (2.10.4)$$

where the α_i are the remaining Feynman parameters, C_0 and D_0 are the Feynman functions of the contracted diagram fig. 2.10.3, and the $N_{\mu\nu}$ are the numerator factors corresponding to the current couplings which we do not write out explicitly. We need the imaginary part of this amplitude in order to calculate the deep inelastic cross-section. This is obtained from (2.10.4) by the substitution

$$1/(g/C_0 - \omega^{-1}) \to -\pi\delta(g/C_0 - \omega^{-1}). \qquad (2.10.5)$$

It is easy to see from the rules of chapter 1 that (2.10.5) imposes the condi-

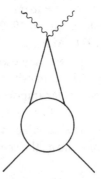

FIGURE 2.10.3 The contracted diagram corresponding to fig. 2.10.2 with $\alpha = 0$.

tion $|\omega| \leqslant |$ and that for a planar diagram, $\omega \geqslant 1$. The physically accessible kinematic region is just $\omega \geqslant 1$.

The fact that (2.10.4) gives a behaviour which is a single power of v^{-1} times a function of ω is the celebrated scaling law first stated by Bjorken (1969). See the discussion in section 3.2 for further details.

If we multiply numerator and denominator in (2.10.4) by ω, the coefficient of ω in the denominator is proportional to g. This implies that the large ω behaviour is controlled by the region of Feynman parameters for which $g = 0$. Since αg is the coefficient of s for fig. 2.10.2 this implies that the larger ω limit is directly related to the large s Regge limit of the diagram. This notion is physically plausible, since $\omega \gg 1$ implies $s \gg |q^2|$. We return to this point in section 3.2.

The simple behaviour found in super-renormalisable ϕ^3 theory is modified when we have a renormalisable theory. The contracted vertex part of fig. 2.10.3 in now logarithmically divergent, signalling that singular configuration effects need to be taken into account which will enhance the asymptotic behaviour accordingly. We discuss some of these effects in section 3.3. The resulting series of powers of $\ln v$ only exponentiates when we consider certain moment integrals over ω, rather than fixed values of ω.

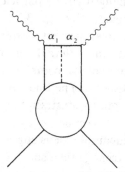

FIGURE 2.10.4 A non-handbag diagram giving a leading contribution in the Bjorken limit. The broken line is a vector particle.

A further complication occurs when vector particles are present in the theory for it is no longer the case that the dominant contribution comes solely from the handbag diagrams. In fig. 2.10.4, although it is necessary to put both α_1 and α_2 to zero to get the end point behaviour the resulting v^{-2} is compensated by the appearance of a factor v in the numerator arising from the exchange of the vector particle. It is possible to attach an arbitrary number of vector particles to the line joining the current vertices without losing the net v^{-1} scaling behaviour.

3
Non-perturbative methods

3.1 Introduction

Calculations based on perturbation theory provide valuable insight but they have a number of drawbacks. Firstly, Regge poles can only be represented by sums of ladders. In elaborate diagrams, such as the Mandelstam diagram of fig. 2.7.2, this considerably complicates the calculation. In the approach on which this chapter is based such difficulties are avoided by using the hybrid, largely non-perturbative, models first introduced by Gribov (1968). They correspond to diagrams containing bubbles or other symbols, representing the (non-perturbative) amplitudes for subprocesses, joined together by lines which represent external particles and exchanged virtual particles (just as in perturbation theory). Examples are provided by figs. 3.2.1, 3.7.4. The internal lines which appear in these diagrams represent the quarks or other fundamental constituents out of which the external hadrons are composed. The bubbles or wavy lines represent the complete amplitude for the mutual interaction of these constituents. The kinematics of the overall process under consideration turns out to constrain these constituent interactions to take place in well-defined regimes, such as, for example, the Regge limit. In this case, if the subamplitude is supposed to behave in the way specified by Regge pole exchange then such an assumption provides the way in which Regge poles can be represented in the calculation without recourse to the ladders of perturbation theory. Not only is this non-perturbative representation of Regge poles easier to calculate with; it also provides a more convincing and general treatment.

A second drawback of the perturbation theory approach is that the very powerfulness of the mathematical techniques provided by Mellin transforms of Feynman integrals sometimes presents the answer with an economy which obscures some of the underlying physics. For example, the reason why two-dimensional integrals are associated with fig. 2.1.2 in the calculation of Regge trajectories is not made very clear by the formalism of the preceding chapter. An enhanced understanding can be

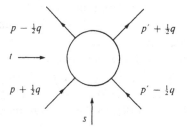

FIGURE 3.1.1 Equal-mass scattering kinematics.

obtained by the momentum–space methods we are about to describe. The basic technical trick is the use of Sudakov (1956) parameters.

As an example consider the process of fig. 3.1.1 for equal-mass particle scattering. The momenta are parametrised as shown and it follows that

$$\left. \begin{aligned} & p{\cdot}q = p'{\cdot}q = 0, \\ & p^2 = p'^2 = m^2 - \tfrac{1}{4}q^2 \equiv \tau, \\ & t = q^2, s = (p + p')^2. \end{aligned} \right\} \tag{3.1.1}$$

In the Regge limit, $s \to \infty$, t fixed, the vectors p and p' will have large components in the overall centre of mass and the vector q, which is space-like and lies in the two-dimensional subspace perpendicular to p and p', will have finite components. Thus asymptotically the external momenta almost lie in the two-dimensional longitudinal subspace defined by p and p'. At high energies there is considerable uncoupling between the longitudinal degrees of freedom in the plane of p and p' and the transverse degrees of freedom in the perpendicular plane containing q. It proves convenient, therefore, to express vectors appearing in the problem in the standard form

$$k = xp + yp' + \kappa, \quad \kappa{\cdot}p = \kappa{\cdot}p' = 0, \tag{3.1.2}$$

which separates the longitudinal degrees of freedom, corresponding to x and y in (3.1.2), from the transverse degrees, corresponding to κ (which is spacelike and effectively two-dimensional since it must lie in the transverse space). Instead of integrating over the four components of k one integrates over x, y, κ. In the limit $s \to \infty$, t fixed, we find

$$\mathrm{d}^4 k \to \tfrac{1}{2} s \, \mathrm{d}x \, \mathrm{d}y \, \mathrm{d}^2 \kappa. \tag{3.1.3}$$

As we shall see in section 3.5 the two-dimensional integrals in fig. 2.1.2 are just integrals over the transverse components κ.

A Sudakov type of parametrisation exists whenever one decides to expand vectors as linear combinations of certain important external vectors which have large components in the high energy regime under consideration, together with such transverse vectors as are necessary to complete the set. The sections which follow will provide a number of examples of the procedure. A particularly useful variant is provided by the parametrisation of (3.4.2) and (3.4.3), suitable for the case of a constituent vector k which is almost parallel to the large momentum p of its parent hadron.

In the succeeding sections we shall discuss the application of these Gribov models of deep inelastic leptoproduction, the production of hadronic particles at large transverse momentum, and Regge theory. The versatility of the technique is well illustrated by its utility in these different physical regimes. A common feature of all its applications is the assumption that constituent subamplitudes decrease very rapidly when the constituent virtual masses become large. It is this condition which restricts the asymptotically significant region of integration over the Sudakov parameters to one in which the subscattering processes represented by the bubbles are evaluated in well-defined kinematic limits. The consequences of relaxing somewhat this condition are discussed in section 3.3.

3.2 Deep inelastic processes

The Gribov hydrid models provide a very natural way (Landshoff *et al.*, 1971) of formulating the *parton model* (see Feynman, 1972) of deep inelastic processes.

As in section 2.10 we are considering the scattering of a virtual photon of momentum q off a target hadron of momentum p, in the Bjorken limit

$$\left. \begin{array}{l} |q^2| \to \infty, \quad \nu = p \cdot q \to \infty, \\ 2\nu / - q^2 = \omega, \quad \text{fixed.} \end{array} \right\} \tag{3.2.1}$$

We may expect that the *handbag diagram* of fig. 3.2.1 will play a dominant role since the single propagator joining the two current vertices will provide the shortest possible path by which the large momentum carried by the virtual photon can propagate through the diagram. (We shall discuss other diagrams later.) The internal lines of the figure represent the constituents or partons out of which the initial hadron is composed. Fig. 3.2.1 appears similar to fig. 2.10.2 but its status is different. In fig. 2.10.2 the lower bubble stands for a particular Feynman subdiagram or set of such diagrams. In fig. 3.2.1 T is the complete parton–hadron

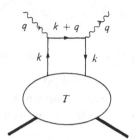

FIGURE 3.2.1 The handbag diagram.

amplitude. If we were to think of it in terms of Feynman diagrams it would be the sum of the infinite set of all possible subdiagrams. However it is not necessary or indeed helpful, to think of T in this way. Instead we shall just assume that it has the properties expected of a hadronic amplitude (for example, Regge behaviour in the appropriate regime) together with the special parton property of rapid decrease when the virtual parton masses become large. A more specific condition on this rate of decrease is stated after equation (3.2.32).

The cross-section will be calculated by the optical theorem from the imaginary part of the amplitude associated with fig. 3.2.1. In a realistic theory there will be a variety of types of partons (the quarks and anti-quarks). Each type will have its corresponding handbag diagram and the total cross-section will be an incoherent sum of the contributions from the diagrams of the different types. Of course the quark partons have spin $\frac{1}{2}$ but for simplicity we first consider the case of scalar partons.

Scalar partons

The scalar partons interact with the photon through the vertex corresponding to the current

$$i\phi^+ \overset{\leftrightarrow}{\partial_\mu}\phi. \qquad (3.2.2)$$

(It is convenient to work throughout with bare Heisenberg fields.) The amplitude of fig. 3.2.1 will be

$$M_{\mu\nu} = \frac{i}{(2\pi)^4} \int d^4k \, \Delta_F'((k+q)^2)(2k+q)_\mu(2k+q)_\nu \, T(k,p), \qquad (3.2.3)$$

where Δ_F' is the parton propagator corresponding to the line joining the two current vertices and T is the amplitude associated with the lower bubble. T is a non-amputated amplitude, that is, it contains the propagators of two vertical sides of the handbag. Δ'_F will have the canonical asymptotic behaviour

$$\Delta'_F((k+q)^2) \sim 1/(k+q)^2, |(k+q)|^2 \to \infty. \qquad (3.2.4)$$

The normalisation of (3.2.4) is determined by the equal-time commutation relations of the ϕ-field. This can be shown by the methods of Bjorken (1966) and Johnson & Low (1966) or by using the Källén–Lehmann representation for Δ'_F:

$$\Delta'_F(x) = \int d\tilde{\mu}^2 \rho(\tilde{\mu}^2) \Delta_F^{(\tilde{\mu}^2)}(x). \qquad (3.2.5)$$

Here $\Delta_F^{(\tilde{\mu}^2)}$ is the ordinary Feynman propagator for mass $\tilde{\mu}^2$. The vacuum expectation value of the equal-time commutator of $\phi(x,0)$ and $\phi(0,0)$ gives the condition

$$\partial \Delta'_F / \partial x_0 \big|_{x_0=0} = \delta^3(x), \qquad (3.2.6)$$

which in turn implies

$$\int d\tilde{\mu}^2 \rho(\tilde{\mu}^2) = 1. \qquad (3.2.7)$$

Equation (3.2.7) and the Fourier transform of (3.2.5),

$$\Delta'_F((k+q)^2) = \int \frac{d\tilde{\mu}^2 \rho(\tilde{\mu}^2)}{(k+q)^2 - \tilde{\mu}^2 + i\varepsilon}, \qquad (3.2.8)$$

then give (3.2.4).

The two important momenta in the problem are p and q so that we introduce the Sudakov parameterisation

$$\begin{aligned} k &= xp + yq + \kappa, \\ \kappa \cdot p &= \kappa \cdot q = 0, \end{aligned} \qquad (3.2.9)$$

with κ spacelike and two-dimensional. In the Bjorken limit (3.2.1)

$$d^4 k \to v\, dx\, dy\, d^2\kappa. \qquad (3.2.10)$$

Since T is supposed to decrease rapidly with increasing $|k^2|$ the significant contribution to (3.2.3) is expected to come from regions of integration over the Sudakov parameters in which k^2 is finite. From (3.2.9) we find

$$k^2 = 2vy(x - \omega^{-1}y) + x^2 m^2 + \kappa^2, \qquad (3.2.11)$$

where $p^2 = m^2$. One region of integration in which k^2 is finite is exhibited by writing

$$y = \bar{y}/2v, \qquad (3.2.12)$$

so that

$$k^2 = x\bar{y} + x^2 m^2 + \kappa^2 + O(v^{-1}). \qquad (3.2.13)$$

(There are other regions of integration which also make k^2 finite. We discuss them below and show that in fact only (3.2.12) gives the dominant asymptotic behaviour.) To leading order in v (3.2.3) takes the form

$$M_{\mu\nu} = \frac{i}{4(2\pi)^4} \frac{1}{v} \int \frac{dx d\bar{y} d^2\kappa (2xp + q + \kappa)_\mu (2xp + q + \kappa)_\nu}{x - \omega^{-1}}$$
$$\times T(s', \mu^2). \qquad (3.2.14)$$

We have used (3.2.4) together with the fact that

$$(k + q)^2 \sim 2v(x - \omega^{-1}), \quad v \to \infty. \qquad (3.2.15)$$

The amplitude T associated with the lower bubble of fig. 3.2.1 is a function of the independent variables

$$\left. \begin{aligned} s' &\equiv (p - k)^2 = (x - 1)\bar{y} + (x - 1)^2 m^2 + \kappa^2, \\ \mu^2 &\equiv k^2 = x\bar{y} + x^2 m^2 + \kappa^2. \end{aligned} \right\} \qquad (3.2.16)$$

It has singularities in these variables and in the dependent crossed-channel variable

$$u' \equiv (p + k)^2 = (x + 1)\bar{y} + (x + 1)^2 m^2 + \kappa^2. \qquad (3.2.17)$$

The $i\varepsilon$ prescription, which we take to hold for all physical amplitudes including the parton–hadron amplitude T, displaces the singularities in s', μ^2, u' below the real axis of the variable in question. Equations (3.2.16) and (3.2.17) tell us what this implies for the sense of displacement of these singularities in terms of the integration variable $\bar{y}$ at fixed x.

If $|x| > 1$ all singularities will be displaced onto the same side of the real $\bar{y}$ contour of integration. In that case completing the $\bar{y}$-integration by a semi-circle at infinity in the opposite half plane will give zero for the integral. This argument assumes that there is sufficient convergence at $|\bar{y}| = \infty$ for the contour closing to be permitted. Since large $\bar{y}$ implies large μ^2 this is guaranteed by the rapid decrease assumed for T when the partons are far off mass-shell. More subtly, the argument also assumes that there are no complex singularities in $\bar{y}$, associated with higher-order Landau singularities in the variables s', u', μ^2 (Eden et al., 1966, ch. 2), which would also prevent the contour closing to give zero. A study of perturbation theory analytic properties indicates that though such complex singularities exist they are placed harmlessly outside the regions of integration over $\bar{y}$ that we are considering.

If $0 < x < 1$, the u' and μ^2 cuts are displaced to one side of the real $\bar{y}$ axis and the s' cut to the other. It is then possible to wrap the contour round the s' cut, as in fig. 3.2.2, to give a non-zero integral. Similarly, if $-1 < x < 0$, the contour can be wrapped around the u' cut of T.

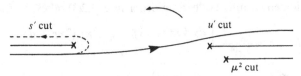

FIGURE 3.2.2 The $\bar{y}$-plane if $0 < x < 1$, showing the contour displacement necessary to enwrap the s'-cut.

Thus T can be replaced in (3.2.13) by

$$-2i \left[\theta(x)\theta(1-x)\operatorname{Im} T_R + \theta(-x)\theta(1+x)\operatorname{Im} T_L \right], \qquad (3.2.18)$$

where the imaginary parts correspond to $(2i)^{-1}$ times the discontinuities across the right hand s' and left hand u' cuts respectively.

The integral given by (3.2.14) and (3.2.18) is manifestly real provided the denominator $(x - \omega^{-1})$ does not vanish. Since $|x| < 1$ this can only happen if $|\omega| > 1$. The $i\varepsilon$ implicit in the propagator (3.2.4) from which the denominator $(x - \omega^{-1})$ derives, then implies that if this denominator vanishes the integral acquires an imaginary part given by the same integral with the substitution

$$1/(x - \omega^{-1}) \rightarrow -\pi\delta(x - \omega^{-1}). \qquad (3.2.19)$$

Equation (3.2.19) states that x, the fraction of the parent hadron momentum p carried by the parton which interacts with the photon, is just equal to ω^{-1}. This exactly agrees with the ideas of the naive parton model based on a simple application of the impulse approximation (see Feynman, 1972). The same interpretation is contained, less transparently, in equation (2.10.5), as can be seen from the rules for the Y_i given in section 1.2.

$\operatorname{Im} M_{\mu\nu}$ is decomposed into a gauge-invariant tensor expansion defining two independent structure functions W_1 and W_2 according to the equation

$$(1/2\pi)\operatorname{Im} M_{\mu\nu} = -(g_{\mu\nu} - q_\mu q_\nu/q^2)W_1(\nu, q^2)$$
$$+ [p_\mu - (p \cdot q/q^2)q_\mu][p_\nu - (p \cdot q/q^2)q_\nu]W_2(\nu, q^2). \qquad (3.2.20)$$

The Bjorken scaling law states that in the limit (3.2.1)

$$W_1(\nu, q^2) \rightarrow F_1(\omega), \quad \nu W_2(\nu, q^2) \rightarrow F_2(\omega), \qquad (3.2.21)$$

with the F_i depending, as indicated, only on the dimensionless Bjorken variable ω. The behaviour (3.2.21) can be read off from (3.2.14), (3.2.18) and (3.2.19). For example, identifying W_2 as the coefficient of $p_\mu p_\nu$ we find

$$\nu W_2 = -\frac{1}{(2\pi)^4 \omega^2} \int d\bar{y} d^2\kappa \, \mathrm{Im} \, T_R(s'\mu^2), \qquad (3.2.22)$$

where (3.2.16) and (3.2.19) show that the right hand side is indeed a function of ω only. In making the identification in this way we should remember that for (3.2.14), in the integration over the two-dimensional vector κ, $\kappa_\mu \kappa_\nu$ is to be replaced by

$$\tfrac{1}{2}\kappa^2 \Delta_{\mu\nu}, \qquad (3.2.23)$$

where $\Delta_{\mu\nu}$ is the diagonal Kronecker-delta matrix in the two-dimensional transverse subspace in which κ lies. A simple piece of algebra enables $\Delta_{\mu\nu}$ to be written in terms of $g_{\mu\nu}$ and p and q as

$$\Delta_{\mu\nu} = \left[g_{\mu\nu} - \frac{p_\mu p_\nu}{m^2 - \nu^2/q^2} - \frac{q_\mu q_\nu}{q^2 - \nu^2/m^2} + \frac{(p_\mu q_\nu + p_\nu q_\mu)\nu}{m^2 q^2 - \nu^2} \right]. \qquad (3.2.24)$$

In fact (3.2.24) does not contribute to the leading behaviour of the coefficient of $p_\mu p_\nu$ in the Bjorken limit since the appropriate term in (3.2.24) vanishes like ν^{-1}. Thus (3.2.11) comes entirely from the explicit ps in (3.2.14).

If W_1 is identified as the coefficient of $-g_{\mu\nu}$ it arises in (3.2.14) solely from (3.2.24) and vanishes like ν^{-1}. Thus W_1 satisfies (3.2.21) for scalar partons with the trivial value

$$F_1 = 0. \qquad (3.2.25)$$

A non-trivial F_1 will be found when we consider spin $\tfrac{1}{2}$ partons.

Straightforward algebra also shows that (3.2.19) substituted into (3.2.14) ensures that W_2 appears multiplied by the correct gauge-invariant combination of ps and qs specified by (3.2.20). Although the handbag diagram is not gauge invariant by itself we shall show later that in the Bjorken limit it dominates the other diagrams associated with it in a gauge-invariant set, so that to leading order it must by itself give a gauge-invariant contribution.

It is now necessary to consider regions of integration other than that given by (3.2.12) which also make k^2 finite. One could, for instance, write

$$y = x\omega + \bar{y}/2\nu. \qquad (3.2.26)$$

However in this case s' is large and independent of $\bar{y}$,

$$s' \sim -2x\omega v. \tag{3.2.27}$$

Therefore the $\bar{y}$-integration has associated with it only the singularities in μ^2 (which would be present in the coupling function of the Regge poles whose exchange would dominate T in the regime (3.2.27)). It is then always possible to complete the $\bar{y}$ contour at infinity to give zero, so that (3.2.26) does not after all contribute to the leading asymptotic behaviour. Similar arguments can be used to exclude contributions arising from making k^2 finite by cancelling the term in v, by allowing κ^2 to be large and (since κ is spacelike) negative.

It is very characteristic of Sudakov variable calculations that some of the apparently significant regions of integration are found not to contribute to the asymptotic behaviour because in fact they give terms which on completion of a contour can be shown to lead to vanishing coefficients. It is always necessary to be watchful for this in calculations.

Before we leave the handbag diagram it is instructive to consider a slightly different way of evaluating its contribution, making use of the Källén–Lehmann representation (3.2.8). (In passing it is interesting to note that Δ'_F does not have to have a particle pole but can have just a continuum contribution; that is, ρ in (3.28) need not contain a δ-function piece. This observation may be relevant to the question of quark confinement (see also, Polkinghorne, 1975).)

The substitution

$$x = \omega^{-1} + \bar{x}/2v \tag{3.2.28}$$

makes $(k+q)^2$ finite:

$$(k+q)^2 = \bar{x} + \omega^{-2}m^2 + \kappa^2, \tag{3.2.29}$$

and this is the only variable in which $\bar{x}$ appears. Taking the imaginary part replaces (3.2.8) by

$$-\pi\rho((k+q)^2). \tag{3.2.30}$$

The integration over $\bar{x}$ factors out from the rest of the integral and from (3.2.29) we see that it is equivalent to integrating over $\tilde{\mu}^2 = (k+q)^2$. The range of $\bar{x}$ is necessarily restricted to give positive values of $\tilde{\mu}^2$ since only the latter can contribute to the imaginary part. From (3.2.7) and (3.2.30) this integral just gives a factor $-\pi$. A factor of $(2v)^{-1}$ arises from the change of variable $dx \to d\bar{x}/2v$, so that we see that this way of calculating also leads to (3.2.22). One can summarise this method by saying that (3.2.22) and (3.2.28) together make both k^2 and $(k+q)^2$ finite at a net cost of a $(2\nu)^{-1}$ behaviour in the amplitude (taking account of (3.2.10)).

Other diagrams

It is necessary to consider whether further significant contributions can arise from diagrams which are not of the handbag type. Thus we consider fig. 3.2.3, conventionally called the 'cats-ears' diagram, in which T_6 is a connected four-parton two-hadron amplitude. The gauge-covariant partners of the handbag, obtained by moving the current vertices along the charged particle lines, will be of cats-ears type.

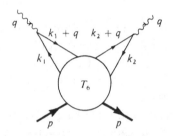

FIGURE 3.2.3 The cats-ears diagram.

There are now two independent parton momenta, k_1 and k_2, and we need to make $k_1^2, (k_1 + q)^2, k_2^2, (k_2 + q)^2$ finite. In analogy with the second method of calculating the handbag diagram we see that making each pair finite costs a factor of $(2\nu)^{-1}$. Thus the net behaviour of fig. 3.2.3 must be at least ν^{-1} below the scaling behaviour unless T_6 contains within itself a compensating factor of ν. One case in which this happens is if there are point-coupled vector particles in the theory. The discussion of fig. 2.10.4 illustrates this. A general treatment of vector particles using the parton model has been given by Nash (1973). The extra contributions that they produce are rather trivially interpretable in a formalism based on a bilocal lightcone algebra (Fritzsch & Gell-Mann, 1971) as due to gauge-invariance properties of the vector mesons.

One might have supposed that an alternative source of a factor of ν would have been the presence of an exchanged Pomeron within T_6. While counting powers of ν would make such a term significant, in fact it can be shown by a contour closing argument (Landshoff & Polkinghorne, 1971b) that the coefficient of the leading behaviour vanishes. This is an example of the cancellation of the effects of final state interactions. A detailed discussion is given in section 4.2.

Limiting behaviour of structure functions

It is useful to rewrite (3.2.22) in terms of the natural variables s' and μ^2 so that it becomes

$$\nu W_2 = \frac{1}{2(2\pi)^3} \frac{1}{\omega(\omega - 1)} \int \mathrm{d}s' \, \mathrm{d}\kappa^2 \, \mathrm{Im} \, T_{\mathrm{R}}(s', \mu^2), \qquad (3.2.31)$$

with

$$\mu^2 = \frac{s'}{1 - \omega} + \frac{m^2}{\omega} + \kappa^2 \frac{\omega}{\omega - 1}. \qquad (3.2.32)$$

Since $\kappa^2 \leqslant 0$, $\omega \geqslant 1$, and s' must be positive, (3.2.32) implies that μ^2 is negative throughout most of the region of integration. This contrasts with the impulse approximation of the naive parton model which (incorrectly) supposes the partons to be 'on mass-shell' before and after their interaction with the current.

The decrease of $\mathrm{Im} \, T_{\mathrm{R}}$ with increasing $|\mu^2|$ provides the convergence factor which makes the integral (3.2.31) well defined. This requirement is the precise definition of the 'softness', or rapid decrease, in the off-shell behaviour of parton amplitudes which is necessary for the application of the ideas of this section. Effectively this means that the range in $|\mu^2|$ is limited. Through (3.2.32) this imposes restrictions on the ranges of s' and κ^2. As ω increases the permitted range in s' increases so that large values of s' become increasingly important. This is why the large ω behaviour of νW_2 is related to (large s') Regge behaviour, as we noted in section 2.10. A formal demonstration of this fact can be given by changing the variable in (3.2.22) from $\bar{y}$ to

$$z = -\omega^{-1}\bar{y}. \qquad (3.2.33)$$

If $\mathrm{Im} \, T_{\mathrm{R}}$ at large s' and fixed μ^2 behaves like

$$\beta(\mu^2)(s')^{\alpha_0}, \qquad (3.2.34)$$

then taking the large ω limit under the integral sign gives the behaviour

$$\nu W_2 \sim \omega^{\alpha_0 - 1} \frac{1}{2(2\pi)^3} \int_0^\infty \mathrm{d}z \, \mathrm{d}\kappa^2 \, \beta(\kappa^2 - z) z^{\alpha_0}. \qquad (3.2.35)$$

The argument of β is negative and the decrease of $\beta(\mu^2)$ at large $|\mu^2|$ will ensure the convergence of the integral in (3.2.35). Thus we find

$$F_2(\omega) \sim \omega^{\alpha_0 - 1}, \quad \omega \to \infty. \tag{3.2.36}$$

If the Pomeron ($\alpha_0 = 1$) is present $F_2(\omega)$ will tend to a constant at large ω.

It is also possible to discuss the limiting behaviour of $F_2(\omega)$ at the threshold end of its kinematic range, where $\omega \to 1$. In this case (3.2.36) implies that μ^2 becomes large throughout the whole region of integration. If T were to have the simple behaviour

$$\text{Im } T_R \sim (\mu^2)^{-\gamma} f(s'), \quad |\mu^2| \to \infty, \quad s' \text{ fixed}, \tag{3.2.37}$$

then the threshold behaviour would be

$$F_2(\omega) \sim (\omega - 1)^{\gamma - 1}, \quad \omega \to 1, \tag{3.2.38}$$

if we assume the integral

$$\int_{s_0'}^{\infty} ds' \int_{-\infty}^{0} d\kappa^2 (s' - \kappa^2)^{-\gamma} f(s') = (\gamma - 1)^{-1} \int_{s_0'}^{\infty} ds' (s')^{1 - \gamma} f(s'), \tag{3.2.39}$$

to be convergent. It is a popular conjecture, supported by some specific models (Drell & Yan, 1970a; West 1970; Landshoff & Polkinghorne, 1973a), that γ is related to the behaviour of the electromagnetic form factor at large q^2,

$$F(q^2) \sim |q^2|^{-p}, \tag{3.2.40}$$

by the Drell–Yan–West relation

$$\gamma = 2p. \tag{3.2.41}$$

An investigation of form factor behaviour by means of Sudakov parameters (Landshoff et al., 1971) makes it clear that (3.2.41) can not be a general result. If the parton analogue of fig. 2.10.3 is evaluated in the $|q^2| \to \infty$ limit it is indeed the case that the parton μ^2s associated with the internal lines become large. However the T amplitude associated with the lower bubble of fig. 2.10.3 not only has large μ^2 but also a large momentum transfer $t' = q^2$. In contrast the T amplitude of fig. 3.2.1 is evaluated with momentum transfer $t' = 0$. Only in specific cases (characterised by T being dominated by the exchange of a single 'core' system in the s'-channel, so that its t'-dependence is trivial) could the large μ^2 behaviour of the amplitude be expected to be independent of t'. Only in these cases, therefore, could (3.2.41) be expected to hold.

Spin $\frac{1}{2}$ partons

We now turn to the realistic case of spin $\frac{1}{2}$ partons. The propagator is assumed to have the asymptotic behaviour

$$S'_F(k+q) \sim \gamma \cdot (k+q)/(k+q)^2, \quad |(k+q)^2| \to \infty, \qquad (3.2.42)$$

in place of (3.2.4). Again the normalisation is determined by the equal-time commutation relations. The integrand of (3.2.3) is replaced by

$$\text{tr}[\gamma_\mu S'_F(k+q)\gamma_\nu T]. \qquad (3.2.43)$$

The Sudakov parametrisation (3.2.9), (3.2.12), isolates the significant region of integration.

The amplitude T is now a matrix in the space of the parton spin indices, the trace in (3.2.43) being taken over these indices. We expand T in terms of γ-matrices operating in this parton spin space

$$T = 1 \cdot T^{(1)} + \gamma^\mu T_\mu^{(2)} + \sigma^{\mu\nu} T_{\mu\nu}^{(3)} + \gamma_5 \gamma^\mu T_\mu^{(4)} + \gamma_5 \cdot T^{(5)}, \qquad (3.2.44)$$

where the coefficients $T^{(i)}$ are to be constructed out of the momenta p and k and the hadron spin variables. They are required by Lorentz invariance to have the tensor characters indicated. Substituting (3.2.42) into the trace (3.2.43) shows that the terms in (3.2.44) with an even number of γs give zero. If we average over hadron spins $T_\mu^{(4)}$ is zero, since an axial vector cannot be constructed from p and k alone. (The case of a polarised hadron target, for which $T_\mu^{(4)}$ is non-zero has been discussed by Nash (1971).)

Thus only $T_\mu^{(2)}$ plays a role in the calculation of unpolarised deep inelastic scattering. On invariance grounds it may be written in the form

$$T_\mu^{(2)} = T_1^{(2)} p_\mu + T_2^{(2)} k_\mu, \qquad (3.2.45)$$

with $T_1^{(2)}$ and $T_2^{(2)}$ scalar functions of s' and μ^2. Use of (3.2.9), (3.2.12) and (3.2.19) together with (3.2.37) and (3.2.45), evaluates the contribution of (3.2.38) to the integrand of the imaginary part of $M_{\mu\nu}$ as being

$$-2i(2\nu)^{-1} \cdot 4[(\omega^{-1}p+q)_\mu p_\nu - g_{\mu\nu}(\omega^{-1}p+q)\cdot p$$
$$+ p_\mu(\omega^{-1}p+q)_\nu] \text{Im}(T_{1R}^{(2)} + \omega^{-1} T_{2R}^{(2)}). \qquad (3.2.46)$$

Identifying W_2 by the coefficient of $p_\mu p_2$ we find that νW_2 scales with $F_2(\omega)$ given by (3.2.22) with $\omega^{-2} \text{Im} T_R$ replaced by $2\omega^{-1} \text{Im} [T_{1R}^{(2)} + \omega^{-1} T_{2R}^{(2)}]$. Identifying W_1 by the coefficient of $-g_{\mu\nu}$ we find that it scales and that $F_1(\omega)$ satisfies that Callan–Gross relation

$$2F_1(\omega) = \omega F_2(\omega). \qquad (3.2.47)$$

It is straightforward to check that (3.2.46) gives to leading order the full gauge-invariant tensor decomposition of $M_{\mu\nu}$ in terms of W_1 and W_2 as specified by (3.2.20).

So far we have considered only an electromagnetic current with its γ_μ vertex. It is, of course, equally possible to consider deep inelastic processes in which the probe is a weak interaction with its $\gamma_\mu(1-\gamma_5)$ vertex. It is convenient to consider the latter as a sum of a V-vertex γ_μ and A-vertex $\gamma_\mu\gamma_5$. A handbag with $2V$-vertices or $2A$-vertices gives exactly the same contributions to W_1 and νW_2 that we have already discussed. A handbag with one A-vertex and one V-vertex gives a parity-non-conserving amplitude whose imaginary part is a contribution to W_3, a new structure function which is the coefficient of the parity-odd tensor

$$\tfrac{1}{2}i\varepsilon_{\mu\nu\alpha\beta}\,q^\alpha p^\beta, \qquad (3.2.48)$$

now permitted in the generalisation of (3.2.20)* The identity

$$\mathrm{Tr}\left[\gamma_5\,\gamma_\mu\,\gamma\cdot a\,\gamma_\nu\,\gamma\cdot b\right] = 4\,i\varepsilon_{\mu\nu\alpha\beta}\,a^\alpha b^\beta \qquad (3.2.49)$$

permits the evaluation of the contribution of an AV handbag to W_3 by the same methods used to give the AA and VV contributions to W_1 and W_2. We find that W_3 scales

$$\nu W_3 \to F_3(\omega) \qquad (3.2.50)$$

in the Bjorken limit, and that

$$F_3(\omega) = -2F_1(\omega). \qquad (3.2.51)$$

However, in contrast to (3.2.47), equation (3.2.51) does not give a universal relationship. This is because every parton handbag like fig. 3.2.1 must have added to it the antiparton handbag obtained by reversing the arrows on the parton lines. Such a reversal of lines is equivalent to charge conjugation under which A is even and V odd. Thus the AA and VV antiparton handbag contributions are just obtained from the contributions of parton handbags by replacing the parton amplitude T by the antiparton amplitude $\bar{T}$. The relationship (3.2.47) is clearly preserved by this operation and so holds for any combination of parton and antiparton contributions. However the sign of the AV handbag changes under change conjugation. Therefore F_3 changes sign and so the relation (3.2.51) holds with reversed

---

* The weak interaction currents are not strictly conserved so that further terms should be included in the tensor expansion. However if we neglect the masses of the leptons which act as the carriers of the weak current it is possible to use (3.2.20) with the sole addition of (3.2.48).

sign for antipartons. Thus when parton and antiparton contributions are added together there is no longer a universal relationship between F_1 and F_3.

Sum rules

Consider the contribution of a parton to a vector form factor evaluated at $t = 0$, as illustrated in fig. 3.2.4. We may evaluate this diagram using the methods employed for fig. 3.2.1. Of course fig. 3.2.4 contains no vector q but we can introduce it via (3.2.9) simply as a convenient mathematical artifice in parametrising the vector k. We are now evaluating an amplitude rather than an imaginary part so that x will not be set equal to ω^{-1}. Instead it remains to be integrated over. By the arguments given

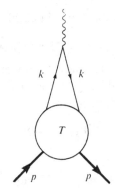

FIGURE 3.2.4 The form factor at $t = 0$.

above, the range of integration is restricted to $|x| \leqslant 1$. Thus we find the form factor corresponding to a parton a can be written as

$$F_a \cdot 2p_\mu, \tag{3.2.52}$$

where

$$F_a = \int_0^1 G_R^a(x)\,dx - \int_{-1}^0 G_L^a(x)\,dx. \tag{3.2.53}$$

G_R and G_L result from wrapping the $\bar{y}$-integration round the s' and u' cuts of T respectively. By crossing, the u' cut of T can be related to right hand s' cut of the antiparton amplitude $\bar{T}$, so that (3.2.53) can be rewritten

$$F_a = \int_0^1 G_R^a(x)\,dx - \int_0^1 G_R^{\bar{a}}(x)\,dx = P_a - P_{\bar{a}}, \tag{3.2.54}$$

where $G^{\bar{a}}$ is the antiparton amplitude, corresponding to reversing the direction of the parton lines in fig. 3.2.4. Equation (3.2.54) expresses the zero momentum vector form factor F_a as the difference of a parton and an antiparton contribution. These contributions count the number of partons a and antipartons $\bar{a}$ in the hadron. The G-functions are simply related to the handbag contributions to νW_2, because we find

$$\left. \begin{aligned} F_2^a(\omega) &= \int \mathrm{d}x \, \delta(x - \omega^{-1}) x \, G_R^a(x), \\ F_2^{\bar{a}}(\omega) &= \int \mathrm{d}x \, \delta(x - \omega^{-1}) x \, G_R^{\bar{a}}(x). \end{aligned} \right\} \tag{3.2.55}$$

Thus the contributions in (3.2.54) can be rewritten as integrals over structure function contributions,

$$\left. \begin{aligned} P_a &= \int_1^\infty \frac{\mathrm{d}\omega}{\omega} F_2^a(\omega), \\ P_{\bar{a}} &= \int_1^\infty \frac{\mathrm{d}\omega}{\omega} F_2^{\bar{a}}(\omega). \end{aligned} \right\} \tag{3.2.56}$$

With Pomeron exchange giving $F_2 \sim$ constant at large ω the integrals in (3.2.56) are logarithmically divergent. The infinities in (3.2.56) then reflect the infinite number of partons and antipartons in the so-called 'sea' associated with the Pomeron (see, for example, Landshoff & Polkinghorne, 1971 a). However, experimentally meaningful sumrules can be obtained from (3.2.56) if combinations of P_a and $P_{\bar{a}}$ are formed in which these infinities cancel. (Since the charge conjugation even Pomeron couples symmetrically to a and $\bar{a}$ these logarithmic divergences are absent in $P_a - P_{\bar{a}}$, as is necessary for (3.2.54) to be well defined. $P_a - P_{\bar{a}}$ is just the number of 'valence' quarks in the hadron.)

An example of this is provided by the *Adler sum rule* (Adler, 1966). Consider neutrino scattering off a proton in a conventional quark theory with the Cabibbo angle of the weak current set equal to zero. In this case only u and d quarks are involved. We may write the structure functions F_2 in terms of quark contributions F_2^a as

$$\left. \begin{aligned} F_2^{\nu P} &= 2(F_2^{\mathrm{d}} + F_2^{\bar{u}}), \\ F_2^{\bar{\nu} P} &= 2(F_2^{u} + F_2^{\bar{d}}), \end{aligned} \right\} \tag{3.2.57}$$

the factors of 2 in (3.2.57) taking account of the existence of both AA and and VV handbags contributing to F_2. The corresponding Ps satisfy the valence quark counting conditions

$$P_u - P_{\bar{u}} = 2, \qquad P_d - P_{\bar{d}} = 1 \tag{3.2.58}$$

for the proton, which behaves like a (uud) state. Then (3.2.56), (3.2.57) and (3.2.58) together give the Adler sum rule

$$\int_1^\infty \frac{d\omega}{\omega}(F_2^{\bar{v}P} - F_2^{vP}) = 2. \tag{3.2.59}$$

Other processes

The techniques which we have used to discuss inclusive deep inelastic scattering can also be used to evaluate amplitudes for other physical processes. Particular interest attaches to these additional applications of parton model ideas since no other ways are known to develop the theoretical discussion of these processes. This contrasts with deep inelastic inclusive cross-sections where the technique of operator product expansions on the lightcone (Cornwall & Jackiw, 1971; Brandt & Preparata, 1971) offers an alternative way of deriving the results obtained above.

One such process is semi-inclusive electroproduction, that is to say the cross-section for the detection of one of the final state particles (Drell & Yan, 1970b; Landshoff & Polkinghorne, 1971b). However the process which has the greatest interest is the *Drell-Yan process* of the hadronic production of heavy muon pairs (Drell & Yan, 1970c; Landshoff & Polkinghorne, 1971b)

$$p + p \rightarrow \mu^+ + \mu^- + \text{anything.} \tag{3.2.60}$$

The regime considered is that in which the energy s and the mass of the pair q^2, are both large, with

$$\tau = q^2/s, \quad \text{fixed.} \tag{3.2.61}$$

The pair is pictured as being produced by the annihilation of a quark from one hadron with an antiquark from the other, the resulting virtual photon materialising as the muons. Fig. 3.2.5 gives the diagram whose imaginary part will yield the cross-section via Mueller's generalised optical theorem.

For fig. 3.2.5 we introduce the natural Sudakov parametrisation

$$\left.\begin{array}{l} k_1 = x_1 p_1 + y_1 p_2 + \kappa_1, \\ k_2 = x_2 p_1 + y_2 p_2 + \kappa_2, \\ \kappa_i \cdot p_1 = \kappa_i \cdot p_2 = 0. \end{array}\right\} \tag{3.2.62}$$

By analogy with our preceding calculations we expect the significant region of integration to be one where $k_1^2, (k_1 + p_1)^2, k_2^2, (k_2 + p_2)^2$, are all finite.

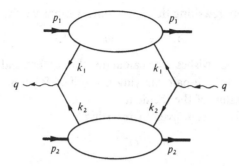

FIGURE 3.2.5 The Drell–Yan process.

This leads to the parametrisation

$$y_1 = \bar{y}_1/s, \quad x_2 = \bar{x}_2/s. \tag{3.2.63}$$

In physical terms this says that the constituent k_i is slowly moving in the rest frame of its parent p_i, and the parameters x_1 and y_2 are the fractions of the parent longitudinal momentum carried by the constituents in a frame where these longitudinal momenta are large. (Another slightly different way of introducing a parametrisation which corresponds to this physical picture is given in equations (3.4.2), (3.4.3).) From (3.2.62), (3.2.63) we find

$$q^2 = (k_1 + k_2)^2 \sim x_1 y_2 s. \tag{3.2.64}$$

The variables $\bar{y}_1, \kappa_1 (\bar{x}_2, \kappa_2)$ only appear in the invariants associated with the top (bottom) bubble of fig. 3.2.5 and the integrals over them just give the appropriate contributions to the structure functions for the quark or antiquark emitted from these bubbles. Thus the cross-section is found to take the scaling form

$$\frac{d\sigma}{dq^2} = \frac{4\pi\alpha^2}{3} \frac{1}{(q^2)^2} W(\tau), \tag{3.2.65}$$

where α is the fine structure constant and $W(\tau)$ is a convolution of quark electroproduction structure functions

$$W(\tau) = \tau^{-1} \sum_a \lambda_a^2 \int_0^\infty d\omega_1 \, d\omega_2 \, \delta(\tau^{-1} - \omega_1\omega_2) F_2^a(\omega_1) F_2^{\bar{a}}(\omega_2). \tag{3.2.66}$$

The sum is taken over quarks of type a and charge λ_a. (Note that no quark charges are included in the F_2 factors; sometimes the Drell–Yan formula (3.2.66) is written with each of the F_2s containing a factor of λ_a^2, in which case there is instead an overall λ_a^{-2}.) The integrals over ω_1

78   NON-PERTURBATIVE METHODS

and ω_2 arise from rewriting the x_1-, y_2-integrations via the identifications

$$\omega_1 = x_1^{-1}, \quad \omega_2 = y_2^{-1}. \tag{3.2.67}$$

Sometimes one wishes to calculate the differential cross-section corresponding to the μ-pair carrying a specified fraction y of the centre of mass momentum of the incident hadrons. From (3.2.62), (3.2.63) and (3.2.67) we see that this is given by inserting

$$\delta(y - \omega_1^{-1} + \omega_2^{-1}) \tag{3.2.68}$$

into (3.2.66).

Annihilation processes

Some interesting features arise from applying the parton model to electron–positron annihilation. We consider first the total cross-section for the process

$$e^+ + e^- \to \text{hadrons}, \tag{3.2.69}$$

at large values of q^2, the square of the centre-of-mass energy. It is necessary to calculate the imaginary part of fig. 3.2.6a which is found to be dominated by the disconnected contribution to the parton–antiparton amplitude, fig. 3.2.6b (Cabibbo, Parisi & Testa, 1970). Since there is only one significant

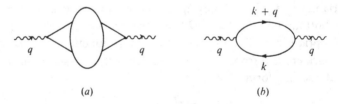

(a)   (b)

FIGURE 3.2.6 Diagrams relating to the total cross-section in e^+e^- annihilation.

momentum in the problem, namely q, we write

$$k = xq + \kappa, \quad \kappa \cdot q = 0, \tag{3.2.70}$$

where κ is a three-dimensional spacelike vector transverse to q. It is necessary to make k^2 and $(k+q)^2$ finite in fig. 3.2.6b, so we write

$$x = -\tfrac{1}{2} + \bar{x}/q^2, \quad \kappa^2 = -\tfrac{1}{4}q^2 + \bar{\kappa}^2, \tag{3.2.71}$$

and the integrals over $\bar{x}$ and $\bar{\kappa}^2$ become integrals over the spectral functions $\rho(\bar{\mu}^2)$ giving the imaginary parts of the parton propagators. The normalisation (3.2.7) then leads to an asymptotic contribution to the total cross-section

$$(\pi\alpha^2/3q^2)\lambda_a^2, \qquad (3.2.72a)$$

for a parton of spin zero and charge λ_a, and

$$(4\pi\alpha^2/3q^2)\lambda_a^2 \qquad (3.2.72b)$$

for a parton of spin $\frac{1}{2}$ and charge λ_a. The extra factor of 4 in (3.2.72b) corresponds to the sum over spin states for the two partons in the loop.

The results (3.2.72) can also be obtained from lightcone arguments at spacelike q^2 supplemented by analytic continuation to timelike q^2. However, only parton model techniques can be used to discuss the inclusive differential cross-section for the production of a final state particle of momentum p:

$$e^+ + e^- \rightarrow p + \text{anything}. \qquad (3.2.73)$$

The leading contribution to the cross-section can be calculated from fig. 3.2.7. The vertical bar denotes that the intermediate states are inserted to calculate the appropriate discontinuity and the $+$ and $-$ labels indicate that the evaluation of a cross-section requires that the right and left hand sides are complex conjugates of each other with $\pm$ iε prescriptions respectively. At first sight fig. 3.2.7 might seem similar to the imaginary part of fig. 3.2.1, which could also be denoted by the insertion of a bar and $\pm$ labels.

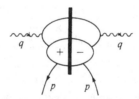

FIGURE 3.2.7 Single-particle production in e^+e^- annihilation.

However the discussion of fig. 3.2.7 is complicated by the fact that q is now timelike so that q^2 is in a region where it has cuts. If the q^2 on the $+$ side is above these cuts, the q^2 on the $-$ side must be in the complex conjugate position below them. In fig. 3.2.1 on the other hand q^2 is negative and free from cuts so that this question does not arise. It is clear, therefore, that fig. 3.2.7 can not be obtained from fig. 3.2.1 by analytic continuation in the single variable q^2, since the q_+^2 has to go along a path above the cuts and the q_-^2 has to go along the complex conjugate path below them.

Annihilation structure functions, $\bar{W}_1$ and $\bar{W}_2$, can be defined for $\omega < 1$, in analogy to the electroproduction structure functions W_1 and W_2 with $\omega > 1$ (for details, see Drell, Levy & Yan, 1970). The argument we have just given makes it clear that $\bar{W}_i$ can not in general, be obtained from W_i by analytically continuing from $\omega > 1$ to $\omega < 1$.

These analytic problems manifest themselves if we try to calculate the $\bar{W}_i$ from fig. 3.2.7. by the same methods we used to calculate the W_i from the imaginary part of fig. 3.2.1. In the latter case the μ^2 cuts are far away from the s' threshold round which we wrap the contour to calculate W_i (see fig. 3.2.2). However with $\omega < 1$ for $\bar{W}_i$ it is easy to see that the μ^2 and s' cuts now overlap. Just as there are two distinguishable $q^2_\pm$ which must be disposed in a complex conjugate fashion to give the cross-section, so there are two distinguishable $\mu^2_\pm$, corresponding respectively to the right and left hand parton k^2 in fig. 3.2.7. If one is above the contour the other must be below it, with the contour wrapped round the s' cut alone, since this gives the discontinuity which corresponds to the cross-section. Therefore fig. 3.2.2 is replaced by fig. 3.2.8. Clearly the contours of these two figures are not in general connected by any analytic continuation in ω since this would treat the two μ^2 (implicitly combined in fig. 3.2.2) in the same way putting both either above or below the contour wrapped round the s' cut.

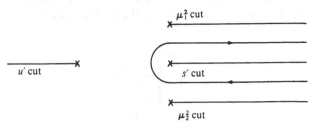

FIGURE 3.2.8 The contour of integration for evaluating $\bar{W}_i$.

Nevertheless some very special cases are known in which such analytic continuation is possible. More significantly there are general grounds for expecting F_2 and $\bar{F}_2$ to display the same threshold behaviour as $\omega \to 1$ from their respective sides (Landshoff & Polkinghorne, 1973a; Gatto, Menotti & Vendramin, 1972).

3.3 Models of scale breaking

The model presented in section 3.2 leads to exact Bjorken scaling. However, phenomenologically it seems that the structure functions exhibit some q^2-dependence. F_2 increases with q^2 for large ω and decreases with q^2 for ω near 1. Scale breaking is not unexpected theoretically for we have seen in section 2.10 that only a super-renormalisable field theory like ϕ^3 has a sufficiently softened off-shell behaviour to make (3.2.22) the convergent integral needed for exact scaling. In the diagrams of renormalisable field theories logarithmic scale breaking effects are present. In this section we consider these matters further and introduce a generalisation of the parton model, called the asymptotically free parton model (Polkinghorne, 1976a, 1977), which accommodates such effects.

Sources of scale breaking

In a renormalisable field theory there are three sources of scale breaking:

(i) A violation of scaling arises for the handbag diagram if the integral (3.2.22) fails to converge. Since this is equivalent to the existence of divergences in the vertex part integral of fig. 3.2.4 this effect is easily identified in renormalisable field theories. The vertex parts are logarithmically divergent and this leads to logarithmic breaking of scaling. An example is provided by ladder diagrams such as that of fig. 3.3.1. These give leading logarithmic effects; that is, each power of g^2 (the coupling constant associated with the exchanged gluon) has associated with it a power of $\ln q^2$, since each extra gluon exchanged creates a further logarithmically divergent loop when fig. 3.3.1 is contracted to fig. 3.2.4.

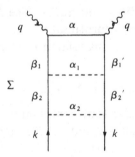

FIGURE 3.3.1 A ladder which will serve as a simple model for A in fig. 3.3.3. The Feynman parameters associated with the lines are indicated.

(ii) Violations of scaling can also arise for cats-ears diagrams. Fig. 3.3.2a shows a simple example of one way this can happen. We consider first the case of a scalar gluon exchange.

The numerator factors appearing in the vertex part can be written

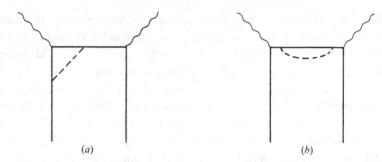

FIGURE 3.3.2 Divergent contributions.

as a sum of a term proportional to k'^2, the symmetric loop momentum (which gives the divergent part of the integral), and a term proportional to the external momenta (which is convergent). We call these the divergent and convergent parts, respectively, and here concentrate on the former. It yields a behaviour proportional to

$$\ln \Lambda^2/M^2, \tag{3.3.1}$$

where Λ^2 is a cut-off parameter, and M a mass scale set by the parameters of the vertex part. If $|q^2|$ is much larger than the other dimensional parameters it will set the scale for M so that in the large q^2 regime (3.3.1) becomes proportional to

$$\ln \Lambda^2/q^2. \tag{3.3.2}$$

This can be verified by direct integration. The Λ^2-dependence is to be removed, according to the renormalisation procedure, by subtracting at a point specified by the renormalisation mass μ^2. At this point (3.3.1) becomes proportional to

$$\ln \Lambda^2/\mu^2. \tag{3.3.3}$$

Thus the renormalised divergent contribution will have a behaviour given by the difference of (3.3.2) and (3.3.3) so that it has the asymptotic behaviour

$$- \ln q^2/\mu^2. \tag{3.3.4}$$

This divergent part, therefore, behaves as though the vertex subdiagram were contracted to a point (thus turning fig. 3.3.2a into an effective Born approximation diagram) together with a scale breaking factor of $\ln q^2$. The reader who finds the discussion given above unduly heuristic can verify the correctness of its conclusion by using the dimensional regularisation procedure outlined in section 2.5 to calculate the renormalised vertex part and then finding its large q^2 behaviour by the methods of chapter 2.

Similar scale breaking effects are associated with self-energy insertions on the line joining the current vertices, as in fig. 3.3.2b. Again one finds an effective Born approximation diagram multiplied by $\ln q^2$. The $\ln q^2$ from the self-energy insertion fig. 3.3.2b cancels exactly against the contribution from one of the two vertex parts of the type of fig. 3.3.2a, leaving a net contribution from only one of the photon vertex parts.

The reason for this is the Ward identity in quantum electrodynamics which requires that Z-factors associated with vertex parts and self-energy parts cancel,

$$Z_{\text{self-energy}} = Z_{\text{vertex}}^{-1}. \tag{3.3.5}$$

Equation (3.3.5) written to order g^2 requires a cancellation of $\ln \Lambda^2$ behaviours, and by (3.3.2) this is linked to a cancellation of the $\ln q^2$ behaviours.

Similar considerations apply to vector gluon exchanges except that the q^2-dependence is distributed between the different insertions in a way which is gauge dependent. The Feynman gauge ($g_{\mu\nu}$ alone in the propagator) gives results for these contributions which are very similar to the scalar case (Stirling, 1978). However, there are axial gauges (Dokshitser, D'Yakonov & Troyan, 1978, Frenkel, Shailer & Taylor, 1979; Pritchard & Stirling, 1980) in which the contributions resulting from type (iii) Feynman gauge terms (discussed in the next paragraph) are incorporated in these type (ii) terms. Of course the net q^2-dependence is gauge independent when a gauge-invariant set of diagrams is considered.

(iii) Finally there are the finite parts of insertions such as fig. 3.3.2. For scalar gluons these do not give scaling contributions but for vector gluons in general they do, as we would expect from the discussion at the end of section 2.10. However the treatment of these finite contributions can be greatly simplified by the choice of a suitable gauge and in the special axial gauges it is believed that all significant contributions are of types (i) and (ii) only (see references given above).

We can summarise the discussion of scale breaking by constructing the generalisation of the handbag diagram to the form show in fig. 3.3.3. A will contain within it all the sources of divergences discussed under (i), (ii) and (iii) above. $\tilde{T}$ is a reduced amplitude for which the integral of (3.2.22) would be convergent. Thus A corresponds to all the hard processes of constituents interacting via quantum chromodynamics and $\tilde{T}$ to the soft wave-function coupling the constituents to the hadrons.

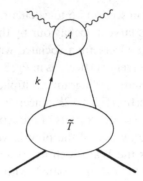

FIGURE 3.3.3 The generalisation of the handbag diagram for the non-scaling case.

In the language of operator product expansions A generates the singular functions in the lightcone and $\tilde{T}$ gives the matrix elements of their operator coefficients between hadron states. Once the behaviour of A is understood the integral over the parton momentum k which couples A to $\tilde{T}$ is performed by the techniques of section 3.2, that is, k is written in the form given by (3.2.9) and (3.2.12).

Ladder exchanges

As an illustration of a contribution to A of type (i) we consider the ladder of fig. 3.3.1, with scalar gluons chosen for simplicity. Each ladder is evaluated in its appropriate Bjorken limit with ω replaced by $x\omega$ since from (3.2.9), (3.2.12),

$$2k \cdot q \sim 2xv \sim x\omega(-q^2).\qquad(3.3.6)$$

This calculation is of the type discussed in chapter 2 and the logarithmic effects arise from the singular configurations associated with the logarithmically divergent loops formed from fig. 3.3.1 by the contraction corresponding to $\alpha = 0$. Inspection shows (Gaisser & Polkinghorne, 1970) that the leading effects are due to terms with the maximum powers of symmetric loop momenta present in numerators of the spin $\frac{1}{2}$ propagators down the sides of fig. 3.3.1. The same leading-order behaviour therefore results if these numerator factors are omitted and their effect simulated by integrating over *six*-dimensional loop momenta. The momentum factors which produce the tensor expansion (3.2.20) come from $\tilde{T}$ (via (3.2.45)) and from the external momenta Y in the numerator of the propagator joining the two current vertices. Once the tensor factors (3.2.20) have been extracted the residual integral resembles the scalar case of section 2.10 but with C^{-2} replaced by C^{-3}. The singular configuration effects are exhibited by the nested sequence of scalings: loops 1 to n, loops 1 to $n-1, \ldots$ loops 1 and 2, loop 1, α. The contribution of the n-loop ladder to νW_2 is then proportional to

$$\left(\frac{g^2}{16\pi^2}\right)^n \int \prod_{i=1}^{n} d\bar{\alpha}_i d\bar{\beta}_i d\bar{\beta}_i' \delta(1 - \bar{\alpha}_i - \bar{\beta}_i - \bar{\beta}_i')$$

$$\times x \prod \bar{\alpha}_i \delta\left(\prod_{i=1}^{n} \bar{\alpha}_i - x^{-1}\omega^{-1}\right) \frac{(\ln q^2)^n}{n!}. \qquad (3.3.7)$$

The presence of the final δ-function prevents the series over n being summed as it stands. However a series which exponentiates is obtained if one considers the moments of νW_2,

$$M_r \equiv \int_1^\infty d\omega \, \omega^{-r} \cdot \nu W_2, \qquad (3.3.8)$$

for which (3.2.7) summed over n gives the behaviour

$$M_r \sim C_r (q^2)^{-a_r}, \qquad (3.3.9)$$

with

$$a_r = - (g^2/16\pi^2)\left[1/r(r+1)\right], \qquad (3.3.10)$$

and

$$C_r = \frac{1}{(2\pi)^3} \int \frac{dx \, ds' \, d\kappa^2}{(1-x)} x^r \text{Im}\left[\tilde{T}_{1R}^{(2)} + x \tilde{T}_{2R}^{(2)}\right]. \qquad (3.3.11)$$

In deriving this result we have used the elementary identity

$$\int d\bar{\alpha} \, d\bar{\beta} \, d\bar{\beta}' \delta(1 - \bar{\alpha} - \bar{\beta} - \bar{\beta}')\bar{\alpha}^{r-1} = \frac{1}{r(r+1)}. \qquad (3.3.12)$$

Momentum space calculation

It is instructive to repeat the calculation of fig. 3.3.1 using the momentum space techniques introduced by Gribov & Lipatov (1972). Let the momentum in the line with parameter β_i be k_i. Instead of expanding k_i in terms of q and p it is convenient to introduce the vectors

$$p' = p - (p^2/2p \cdot q)q, \quad q' = q + (1/\omega)p', \qquad (3.3.13)$$

which have vanishing squares in the Bjorken limit. This is to make expressions like (3.3.15) only linear in x and $\bar{y}$. We therefore write

$$k_i = x_i p' + (\bar{y}_i/2v)q' + \kappa, \quad \kappa \cdot p' = \kappa \cdot q' = 0, \tag{3.3.14}$$

where κ is, of course, also transverse to p and q. Then the squares of the momenta in the sides of the ladder are given by

$$k_i^2 = x_i \bar{y}_i + \kappa_i^2, \tag{3.3.15}$$

and in the rungs of the ladder below the top rung by

$$(k_i - k_{i+1})^2 = (x_i - x_{i+1})(\bar{y}_i - \bar{y}_{i+1}) + (\kappa_i - \kappa_{i+1})^2. \tag{3.3.16}$$

To avoid the $\bar{y}_i$-integrations giving zero we require

$$x > x_n > x_{n-1} > \ldots > x_1 > 0, \tag{3.3.17}$$

bearing in mind that k can be written like (3.3.14) with x_i replaced by x. The $\bar{y}_i$-integrations can then be performed by wrapping the contour round the poles of the propagators corresponding to (3.3.16). If the κ_i^2 are restricted to the regime

$$\kappa^2 \ll \kappa_n^2 \ll \kappa_{n-1}^2 \ll \ldots \ll \kappa_1^2 \ll q^2, \tag{3.3.18}$$

then this implies that

$$\bar{y} \ll \bar{y}_n \ll \bar{y}_{n-1} \ll \ldots \ll \bar{y}_1 \ll 1, \tag{3.3.19}$$

and the result of taking the residues separates the x_i- and κ_i-integrations. The latter take the form

$$\int^{q^2} \frac{d\kappa_1^2}{k_1^2} \int^{\kappa_1^2} \frac{d\kappa_2^2}{\kappa_2^2} \cdots \int^{\kappa_{n-1}^2} \frac{d\kappa_n^2}{\kappa_n^2} \sim \frac{(\ln q^2)^n}{n!}, \tag{3.3.20}$$

when factors of κ_i^2 from the spinor numerators are taken into account. Thus (3.3.20) exhibits the important fact that the logarithmic scale breaking effects in fig. 3.3.1 are associated with the integrations over transverse momenta. Notice that the significant values of these κ_i^2 are all large but not as large as q^2. It is characteristic of logarithmically divergent integrations that this should be so.

The limitation (3.3.18) is justified by its producing the maximum power of $\ln q^2$ from (3.3.20). Some thought shows that the ranges of integration omitted from (3.3.18) would give subdominant contributions at large q^2.

The x_i-integrations take the form

$$\int^x dx_n(1 - x_n) \int^{x_n} \frac{dx_{n-1}}{x_n^2}(x_n - x_{n-1}) \cdots \int^{x_2} \frac{dx_1}{x_2^2}(x_2 - x_1) \frac{1}{x_1 - \omega^{-1}x^{-1}}. \tag{3.3.21}$$

Replacing the last denominator in (3.3.21) by

$$- \pi\delta(x_1 - \omega^{-1}x^{-1}) \tag{3.3.22}$$

in order to calculate the imaginary part, and making the substitutions

$$\left.\begin{array}{l} x_1 = \bar{\alpha}_1 \ldots \bar{\alpha}_n, \\ x_2 = \bar{\alpha}_2 \ldots \bar{\alpha}_n, \\ \ldots \\ x_n = \bar{\alpha}_n, \end{array}\right\} \tag{3.3.23}$$

reduces (3.3.21) to a form where taking the moments reproduces (3.3.9), (3.3.10).

Other contributions

More complicated type (ii) divergent parts will contain several loops. For scalar gluons in the leading logarithmic approximation these loops will have to form a planar nested sequence within the vertex or self-energy parts, as in fig. 3.3.4. In each case we obtain the Born approximation diagram multiplied by a power of $\ln q^2$. Again there is a cancellation between the self-energy part and one of the two vertex parts. Summing over the number of loops exponentiates the $\ln q^2$ (see Stirling, 1978).

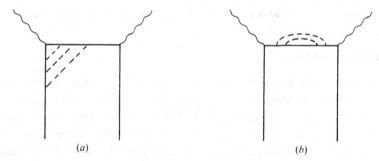

(a)   (b)

FIGURE 3.3.4 Higher-order nested divergent contributions.

The net contribution from the type (ii) divergent parts is found to be

$$(q^2)^{-g^2/32\pi^2}\delta(1 - x^{-1}\omega^{-1}). \tag{3.3.24}$$

The δ-function in (3.3.24) corresponds to the reduced Born approximation term left after the $\ln q^2$ factors have been extracted. Notice that (3.3.24) gives a q^2-dependence for the rth moment which is independent of r.

It is also necessary to consider diagrams which combine type (i) and type (ii) contributions, that is, contain insertions of type fig. 3.3.4 within the ladder fig. 3.3.1. It turns out that all the contributions to the q^2 behaviour of moments combine multiplicatively. This is almost obvious from the way type (ii) terms are extracted but further details can be found in Stirling (1978). Thus the complete leading logarithmic behaviour of the moments of the structure functions in the case of scalar gluons is given by (3.3.9) with the exponent a_r given by

$$a_r = \frac{g^2}{16\pi^2}\left(\frac{1}{2} - \frac{1}{r(r+1)}\right). \tag{3.3.25}$$

The fact that a_r vanishes for $r = 1$, that is, that M_1 scales, is required by general current algebra constraints (see Politzer, 1974).

In the case of vector gluons it is known (Politzer, 1974) that (3.3.25) must be replaced by

$$a_r = \frac{g^2}{16\pi^2}\left[1 - \frac{2}{r(r+1)} + 4\sum_{i=2}^{r}\frac{1}{i}\right]. \tag{3.3.26}$$

If one calculates in the Feynman gauge the first two terms of (3.3.26) arise from the types (i) and (ii) terms. (The extra factor of 2 is due to the gluon's two transverse spin states.) The last term must then arise from the type (iii) contributions. In the axial gauges referred to above it is expected that the whole of (3.3.26) can be obtained from type (i) and (ii) alone. The detailed calculations are of too great complexity to report here.

Asymptotic freedom

In asymptotically free gauge theories (Politzer, 1974), such as quantum chromodynamics, scale breaking is by powers of $\ln q^2$ rather than the powers of q^2 itself exhibited in (3.3.9). This is because the vertex and self-energy parts associated with the rungs and sides of the ladder of fig. 3.3.1 (as the type (ii) terms were associated with the top of the ladder) are expected to modify the coupling g^2 to a running coupling constant $\alpha(k^2)$ which is proportional to $(\ln k^2)^{-1}$ when the momenta k^2 are large. Such an effective coupling constant means that on contracting fig. 3.3.1 the loops of the vertex part are log–log divergent rather than log divergent. This leads to scale breaking by powers of $\ln(\ln q^2)$ in a given ladder, which on summation and exponentiation give an expression similar to (3.3.9) with q^2 replaced by $\ln q^2$.

The detailed verification of this effect from field theory diagrams is clearly immensely complicated. A simple model way of representing the effects is given by the *asymptotically free parton model* (Polkinghorne, 1976a, 1977) which simulates the running coupling constant by replacing the gluon propagators by the behaviour

$$1/k^2 \ln k^2, \quad k^2 \to \infty. \tag{3.3.27}$$

Calculations with (3.3.25) are a little more complicated than with Feynman propagators. It is sometimes (Polkinghorne, 1976b) convenient to introduce a further parameter γ_i for each gluon line and to replace (3.3.27) by the equivalent integral

$$\int_0^\infty \mathrm{d}\gamma_i \frac{1}{(k^2)^{1+\gamma_i}}. \tag{3.3.28}$$

The denominator of (3.3.28) is easily incorporated into symmetric integration by use of the generalised Feynman identity (1.2.7). The leading asymptotic behaviour then comes from a region of integration which includes the end points $\gamma_i = 0$.

The Drell–Yan process

It is of interest to enquire if, in the non-scaling case, (3.2.66) still holds for the Drell–Yan cross-section with the only modification that F_2^a and

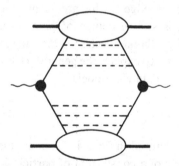

FIGURE 3.3.5 The Drell–Yan process for non-scaling interactions.

$F_2^{\bar{a}}$ are now q^2-dependent. This is called the *Drell-Yan conjecture* (Polkinghorne, 1976c; Kogut, 1976; Hinchcliffe & Llewellyn Smith, 1977). If only type (i) and type (ii) scale breaking processes are present this is indeed manifestly the case. It is necessary to consider fig. 3.3.5 instead of fig. 3.2.5. The black dots represent divergent vertex part contributions.

It is convenient to take the τ-moments of (3.2.67) and obtain

$$W_r^a = M_r^a \cdot M_r^{\bar{a}}, \tag{3.3.29}$$

where

$$W_r^a = \int_0^1 d\tau \, \tau^{r-1} W_a(\tau).$$ (3.3.30)

In fig. 3.3.5 the contributions of the two ladders and the two photon vertices combine multiplicatively in W_r^a. This leads to (3.3.29), since the Ward identity cancellation (3.3.5) means that the net type (ii) term present in F_2 is just that due to one photon vertex and so each M_r^a is the product of one ladder contribution and one photon vertex contribution.

The Drell–Yan conjecture therefore holds in a scalar gluon theory (Stirling, 1978) and is presumed to hold in a vector gluon theory because of the axial gauge structure discussed above.

3.4 Large transverse momentum processes

A further important field of application of the parton model is provided by the production of large transverse momentum particles in hadronic collisions. The basic physical picture is that this is due to the wide angle scattering of constituents present in the initial hadrons, the observed final hadron of large transverse momentum being a fragment of one of the scattered constituents (Berman, Bjorken & Kogut, 1971). The detailed physics, requiring a knowledge of what are the most important constituent scattering processes, is still in course of phenomenological assessment (see Sivers, Brodsky & Blankenbecler, 1976; Jacob & Landshoff, 1978). Our aim in this section is to explain the general parton techniques without commitment to specific detailed models.

Hard scattering

The basic process is pictured in fig. 3.4.1. A constituent of particle 1 with momentum k_1 scatters of a constituent of particle 2 with momentum k_2. We are calculating the inclusive differential cross-section for the production of a final state hadron of momentum p. This is taken to be a fragment of one of the outgoing products of the hard scatter, the latter carrying off momentum k. In some models there is the possibility that the detected hadron is the scattered constituent itself, that is, that $p = k$.

There are three important momenta in the problem, p_1, p_2, p, so that a possible parametrisation of the momenta k_i would be

$$k_i = x_i p_1 + y_i p_2 + z_i p + \chi, \quad \chi \cdot p_1 = \chi \cdot p_2 = \chi \cdot p = 0.$$ (3.4.1)

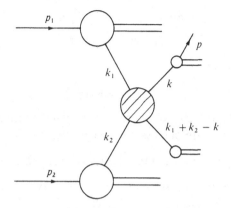

FIGURE 3.4.1 A hard scattering mechanism for the production of particles at large p_T.

χ would be spacelike and confined to the single dimension transverse to p_1, p_2, p. It is possible to develop the theory in this way (Landshoff & Polkinghorne, 1973b, 1974) but a simpler method is to proceed as follows.

Consider all vectors in the overall centre of mass. If the z-axis is in the direction of the vector p_1 this vector can be written as

$$p_1 = \tfrac{1}{2}\sqrt{s}\,(1, 0, 0, 1), \tag{3.4.2}$$

provided we make the approximation, valid in the large s limit, of neglecting the mass of particle 1. We expect that k_1 is almost parallel to p_1 and carries a fraction, x, say, of its longitudinal momentum. It can therefore be written as

$$k_1 = (\tfrac{1}{2}\sqrt{s}\,x_1 + \bar{y}_1/\sqrt{s}, \kappa_1, \tfrac{1}{2}\sqrt{s}\,x_1). \tag{3.4.3}$$

κ_1 is a vector in the transverse plane and it and $\bar{y}_1$ parametrise the deviation of k_1 from parallelism to p_1. We find

$$k_1^2 = x_1 \bar{y}_1 - \kappa_1^2, \tag{3.4.4}$$

$$(p_1 - k_1)^2 = (x_1 - 1)\bar{y}_1 - \kappa_1^2, \tag{3.4.5}$$

and no net factor $\sqrt{s}$ appears in a change of variables from $\mathrm{d}^4 k_1$ to $\mathrm{d}x_1\,\mathrm{d}\bar{y}_1\,\mathrm{d}^2\kappa_1$. In calculating the contribution of fig. 3.4.1 the parameters $\bar{y}_1$ and κ_1 will only appear in the variables k_1^2 and $(p_1 - k_1)^2$ associated with the top bubble. If the constituent is a quark the integral over $\bar{y}_1$ and κ_1 can be identified with the appropriate contribution to the electroproduction structure function $F_2(x_1^{-1})$.

Similar parametrisations, which are rotated and scaled versions of (3.4.2), (3.4.3), can be introduced to relate k_2 to p_2 and k to p. For example the vector p_2 takes the form

$$\tfrac{1}{2}\!\sqrt{s}\,(1,0,0,-1),\qquad\qquad(3.4.6)$$

so that k_2 is written as

$$(\tfrac{1}{2}\!\sqrt{s}\,x_2 + \bar{y}_2/\!\sqrt{s},\kappa_2,-\tfrac{1}{2}\!\sqrt{s}\,x_2).\qquad\qquad(3.4.7)$$

If p corresponds to a particle with transverse momentum p_T emerging at an angle θ in the centre of mass it can be written as

$$R_\theta[(p_T/\sin\theta)(1,0,0,1)] = (p_T/\sin\theta, p_T, 0, p_T\cot\theta),\qquad(3.4.8)$$

where R_θ is a rotation through θ about the y-axis. Then we write k as

$$R_\theta\left[\left(\frac{xp_T}{\sin\theta} + \frac{\sin\theta\,\bar{y}}{2p_T},\kappa,\frac{xp_T}{\sin\theta}\right)\right].\qquad(3.4.9)$$

In each case the transformation of variables costs no power of $\sqrt{s}$ and if the constituents are quarks the integrals over $\bar{y}_2, \kappa_2$ and $\bar{y}, \kappa$ correspond, respectively, to contributions to $F_2(x_2^{-1})$ and to the annihilation fragmentation function $\bar{F}_2(x)$.

The parameters x also appear in the variables associated with the hard scattering process:

$$\left.\begin{aligned}
\hat{s} &\equiv (k_1 + k_2)^2 \sim x_1 x_2 s,\\
\hat{t} &\equiv (k_1 - k)^2 \sim -\tfrac{1}{2}xx_1 x_T \tan\tfrac{1}{2}\theta s,\\
\hat{u} &\equiv (k_2 - k)^2 \sim -\tfrac{1}{2}xx_2 x_T \cot\tfrac{1}{2}\theta s,
\end{aligned}\right\}\qquad(3.4.10)$$

with

$$x_T = 2p_T/\!\sqrt{s}.\qquad\qquad(3.4.11)$$

In addition they appear in the mass associated with the fourth constituent,

$$(k_1 + k_2 - k)^2 \sim s(x_1 x_2 - \tfrac{1}{2}xx_T(x_1 \tan\tfrac{1}{2}\theta + x_2 \cot\tfrac{1}{2}\theta)).\quad(3.4.12)$$

The basic postulate of the parton model requires that (3.4.12) should be made finite. The problem can be solved by a technique similar to that employed in the second method of calculating the handbag diagram described in section 3.2. We write

$$x = x_0 - 2\bar{x}/x_T s(x_1 \tan\tfrac{1}{2}\theta + x_2 \cot\tfrac{1}{2}\theta),\qquad(3.4.13)$$

with x_0 the value that makes (3.4.12) finite,

$$x_0 = \frac{2x_1 x_2}{x_T(x_1 \tan\tfrac{1}{2}\theta + x_2 \cot\tfrac{1}{2}\theta)}.\qquad(3.4.14)$$

Since $\bar{x}$ appears only in the argument of the spectral function associated with the propagator of the fourth constituent its integral factors out and gives a trivial value 1 by (3.2.7).

The flux factor in calculating the cross-section is proportional to s. If the matrix element for the hard scatter behaves in the high energy wide angle limit (3.4.10) like

$$(\hat{s})^{-r}\phi(\hat{t}/\hat{s}),\tag{3.4.15}$$

then the differential cross-section takes the form

$$E\,\mathrm{d}\sigma/\mathrm{d}\boldsymbol{p} \sim s^{-n}f(x_T,\theta),\tag{3.4.16}$$

where

$$n = 2 + 2r.\tag{3.4.17}$$

The function f depends only on the dimensionless variables x_T and θ, and is given by an integral over x_1 and x_2 (with x set equal to x_0) which convolutes $F_2(x_1^{-1}), F_2(x_2^{-1})$, $\bar{F}_2(x_0)$ and ϕ. If the hard scatter is quark–quark scattering mediated by a vector gluon exchange we find $r = 0$. In that case the process is said to be *scale free* since the dimensions then come entirely from the explicit s^{-2} factor in (3.4.16).

Clearly equation (3.4.16) can be rewritten in the equivalent form

$$E\,\mathrm{d}\sigma/\mathrm{d}\boldsymbol{p} \sim (p_T^2)^{-n}\hat{f}(x_T,\theta),\tag{3.4.18}$$

with

$$\hat{f}(x_T,\theta) = (\tfrac{1}{2}x_T)^{2n}f(x_T,\theta).\tag{3.4.19}$$

Exclusive scattering–dimensional counting

The constituent hard scatter at the heart of fig. 3.4.1 is an example of an exclusive process with large momentum transfer. Such processes are directly observable if we consider hadrons in place of constituents. The best studied example experimentally is proton–proton scattering at high energy and wide angle. A simple and phenomenologically successful account of such hadronic exclusive cross-sections is given by the idea of dimensional counting (Brodsky & Farrar, 1975; Matveev, Muradyan & Tavkhelidze, 1973).

Each participating hadron is pictured as composed of the minimum number of constituents (qqq for baryons, $\bar{q}q$ for mesons). These constituents interact with each other through point-coupled vector gluons. The minimum number of interactions take place which are necessary to

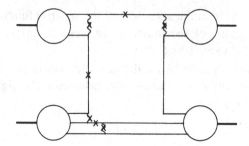

FIGURE 3.4.2 A contribution to dimensional counting.

transfer the large momentum through the system. Fig. 3.4.2 gives an example of such a process for meson–baryon scattering; the lines with crosses carry the large momentum. The corresponding factors in the matrix element are s^{-1} for each gluon propagator carrying a large momentum; $s^{-1/2}$ for each quark propagator carrying large momentum; $s^{1/2}$ for each pair of quark wave-functions connected by an arc of quark lines. Together these factors give a net behaviour for the matrix element of $(s^{-1})^4 \cdot (s^{-1/2})^3 \cdot (s^{1/2})^5 = s^{-3}$, producing an s^{-8} behaviour for $d\sigma/dt$.

In general we see that these rules lead to a cross-section for an exclusive process of the form

$$d\sigma/dt \sim s^{-n}\phi(t/s), \qquad (3.4.20)$$

with the exponent n given by

$$n = \sum n_{\text{constituents}} - 2. \qquad (3.4.21)$$

The total number of participating constituents is to be reckoned according to the quark model values for the initial and final state particles; that is, 3 for each baryon and 2 for each meson. This gives $n = 10$ for proton–proton scattering (for which there is good phenomenological support (Landshoff & Polkinghorne, 1973c))$n = 8$ for pion–proton scattering, etc. Of course (3.4.21) gives the scale-free value $n = 2$ for quark–quark scattering.

These results can also be obtained by the techniques of chapter 2. Applied to fig. 3.4.2 they lead us to identify (3.4.20) (3.4.21) with an end point contribution, the lines with crosses in the figure now being those which are contracted out to give the high energy asymptotic behaviour. However, contained within the hadronic wave-functions of fig. 3.4.2 are further gluon exchanges which serve to bind the quarks to form the hadrons. If these exchanges were exhibited explicitly they would make clear the existence of several d-line scaling sets and there would also be singular configuration modifications to (3.4.20). For asymptotically free theories we might hope that these latter modifications would have a logarithmic rather than power-law character, like the scale breaking effects in deep inelastic scattering. However it has not been proved that this is the case. Related to this is the question of whether soft gluon corrections, like those discussed in section 2.8, might dampen the quark–quark scattering amplitudes below their scale-free values (Polkinghorne, 1974). In this case the analogy with infra-red problems in quantum electrodynamics suggests that this might not be so, provided one includes also the effects associated with a soft gluon cloud around the scattered quark. In this case the quark–quark scattering is no longer strictly exclusive.

A yet more serious problem for dimensional counting is posed by the question of multiple scattering corrections to (3.4.20), to which we now turn.

Multiple scattering mechanisms

Landshoff (1974) considered a multiple scattering process such as that illustrated by fig. 3.4.3 for the case of pion–pion scattering. Each consti-tuent from one of the pions scatters off a constituent drawn from the other pion. The scattered constituents are then constrained to have momenta which are sufficiently closely aligned in pairs for these pairs to recombine to give the final state pions. One might suppose this to be a very unlikely eventuality but it proves not to be so.

The momenta k_i, k_i' are required to be closely aligned with the momenta p_i of their associated hadron. This is achieved by introducing the appro-priate parametrisation of type (3.4.2), (3.4.3) for each k_i. It is then auto-matic that $k_i' = p_i - k_i$ is correctly aligned. No net factors of the large

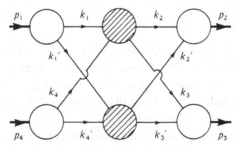

FIGURE 3.4.3 The Landshoff multiple scattering mechanism for pion–pion inter-actions.

variable s arise from this procedure. If we assume as usual that the quark scatterings are scale free their matrix elements also provide no factors of s. The only source of such factors is the δ-function which enforces the internal conservation of momentum,

$$k_1 + k_4 = k_2 + k_3. \tag{3.4.22}$$

Three components of these vectors are large like $\sqrt{s}$ and the fourth, transverse to the scattering plane, is bounded. Thus a factor of

$$s^{-3/2} \tag{3.4.23}$$

arises from these δ-functions. Once (3.4.2) holds, it, together with overall momentum conservation, implies the corresponding relation for the k_i'. Thus (3.4.23) gives the total asymptotic behaviour of the amplitude for fig. 3.4.3, which in turn gives a cross-section of the form (3.4.20) with $n = 5$, in contrast with the dimensional counting value of $n = 6$ for pion–pion scattering.

Proton–proton scattering can be discussed in terms of a similar mechanism with three quark hard scatters. There are then two independent conditions of the type (3.4.22) and thus two factors (3.4.23). The matrix element therefore behaves like s^{-3} and the cross-section like s^{-8}, two powers of s above the s^{-10} of dimensional counting. Ultimately, therefore, the Landshoff multiple scattering process should dominate. However, in the absence of reliable estimates of the relative magnitudes of the ϕ-functions associated with the two processes it is not possible to say at what energies the slower decrease of the Landshoff process will make itself manifest.

The idea of multiple scattering mechanisms has also been applied to the inclusive production of large p_T particles. (Landshoff, Polkinghorne, & Scott, 1975).

3.5 Ladder diagrams

The generation of Regge poles by ladders is in most respects more conveniently treated by the α-space techniques of chapter 2 than by Sudakov parameters. However it is also possible (Halliday & Saunders, 1969) to discuss ladders by using the Sudakov technique and it is of interest to outline how the argument runs.

It is convenient to consider the imaginary part of the ladder amplitude associated with the intermediate state with particles of momenta $k_1, \ldots, k_n$, shown in fig. 3.5.1; that is, each k_i is on its mass-shell, $k_i^2 = m^2$. We write

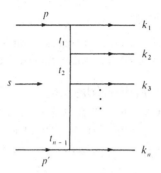

FIGURE 3.5.1 A ladder process.

the k_i in the Sudakov form

$$k_i = x_i p + y_i p' + \kappa_i, \quad \kappa_i \cdot p = \kappa_i \cdot p' = 0. \qquad (3.5.1)$$

Momentum conservation implies that

$$\sum x_i = 1, \quad \sum y_i = 1, \quad \sum \kappa_i = 0, \qquad (3.5.2)$$

and the finite mass conditions for the particles enforce

$$k_i^2 = x_i y_i s + (x_i - y_i)^2 m^2 + \kappa^2 = m^2. \qquad (3.5.3)$$

That is, we require

$$x_i y_i \sim s^{-1}. \qquad (3.5.4)$$

If we choose to satisfy (3.5.4) by making

$$x_i = O(s^{-\gamma_i}), \quad y_i = O(s^{-1+\gamma_i}), \qquad (3.5.5)$$

then a slightly involved but elementary algebraic argument shows that demanding finite values of the momentum transfers t_i down the sides of the ladder requires that

$$0 = \gamma_1 \leqslant \gamma_2 \leqslant \ldots \leqslant \gamma_n = 1. \qquad (3.5.6)$$

This implies that the longitudinal momenta of successive particles in the ladder are ordered in the overall centre of mass. It is convenient to transform to a set of parameters which include the γ_i as variables. The Jacobian of this transformation produces the factors of $\ln s$ which correspond to taking the imaginary part of (2.1.12). The t_i become functions of the κ_i only and it is possible to exhibit the required factor $(K(t))^{n-1}$. The two-dimensional integral defining $K(t)$ is thus understood as arising from an integral over the transverse degrees of freedom of the intermediate state particles.

3.6 Multi-Reggeon exchange

Consider the kinematic regime corresponding to the double Regge exchange diagram of fig. 2.9.2. The final state momenta can be written as

$$\left.\begin{aligned}
p_1' &= x_1 p_1 + (y_1/s)p_2 + \kappa_1,\\
p_2' &= (x_2/s)p_1 + y_2 p_2 + \kappa_2,\\
p_3' &= (x_3/s^\gamma)p_1 + (y_3/s^{1-\gamma})p_2 + \kappa_3,\\
\kappa_i \cdot p_1 &= \kappa_i \cdot p_2 = 0; \quad 0 < \gamma < 1.
\end{aligned}\right\} \tag{3.6.1}$$

Momentum conservation,

$$p_1 + p_2 = p_1' + p_2' + p_3', \tag{3.6.2}$$

then implies that

$$\left.\begin{aligned}
x_1 &= 1 + O(s^{-\gamma}), \quad y_2 = 1 + O(s^{-1+\gamma})\\
\kappa_1 &+ \kappa_2 + \kappa_3 = 0.
\end{aligned}\right\} \tag{3.6.3}$$

The subenergies satisfy

$$\left.\begin{aligned}
s_1 &\equiv (p_1' + p_3')^2 \sim y_3 s^\gamma,\\
s_2 &\equiv (p_2' + p_3')^2 \sim x_3 s^{1-\gamma}.
\end{aligned}\right\} \tag{3.6.4}$$

Therefore

$$s_1 s_2 \sim \eta s, \tag{3.6.5}$$

where

$$\eta = x_3 y_3 = m^2 + \kappa_3^2. \tag{3.6.6}$$

This last equation follows from the finite mass condition $p_3'^2 = m^2$. η is sometimes called the *transverse mass* of particle 3. At high energy η^{-1} is asymptotically a linear function of the cosine of the Toller angle ω (Toller, 1965), defined as the angle between the production planes of the final state particles.

From (3.6.3) the momentum transfers carried by the two Reggeons are given by

$$\left.\begin{aligned}
t_1 &\equiv (p_1 - p_1')^2 = \kappa_1^2,\\
t_2 &\equiv (p_2 - p_2')^2 = \kappa_2^2.
\end{aligned}\right\} \tag{3.6.7}$$

The coupling of the two Reggeons to p_3' will be a function not only of t_1 and t_2 but also of the Toller variable η. It is of interest to investigate the way this coupling depends on η.

The two-Reggeon–particle coupling

The simple model for the coupling is provided by the hybrid model of fig. 3.6.1. The triangle represents Feynman propagators with a circulating loop momentum k. The wavy lines represent subamplitudes, which are functions of $s_1, t_1; s_2, t_2$, respectively, and each is taken to have Regge pole behaviour in the Regge subregimes defined by (3.6.4), (3.6.7). One could parametrise k in the same way as p_3', so that all masses in the triangle loop are finite. The integral evaluated this way (Drummond, 1968) will then give a model for the η-dependence of the coupling. However the answer is found to take a form which can be discussed more illuminatingly in the following way (Drummond, Landshoff, & Zakrzewski, (1969a).

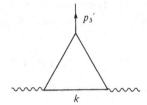

FIGURE 3.6.1 A model for the two-Reggeon one-particle coupling.

In fig. 2.9.2 the variables s, s_1 and s_2 all approach the real axis in the physical limit from the upper half of the complex plane and the amplitude is analytic in these half planes. These facts may be represented by the *Fourier form*

$$\int_0^\infty d\lambda\, d\lambda_1\, d\lambda_2\, \psi(\lambda, \lambda_1, \lambda_2; t_1, t_2)\, e^{i(\lambda s + \lambda_1 s_1 + \lambda_2 s_2)}. \qquad (3.6.8)$$

The positive ranges of integration of the λs in (3.6.8) reflect the upper plane analyticity in the s-variables, for the integral is strongly convergent if any of these variables is given a positive imaginary part. ψ is an arbitrary weight function. If s is expressed in terms of s_1 and s_2 by (3.6.5), and (3.6.8) is then Mellin transformed with respect to s_1 and s_2, the change of variables

$$\lambda_1 s_1 = u_1, \quad \lambda_2 s_2 = u_2, \quad \lambda = \lambda_1 \lambda_2 y, \qquad (3.6.9)$$

enables us to write the Mellin transform as

$$G(\beta_1, \beta_2) = \int_0^\infty dy\, d\lambda_1\, d\lambda_2\, \psi(\lambda_1 \lambda_2 y, \lambda_1, \lambda_2; t_1, t_2)$$

$$\times \lambda_1^{\beta_1 + 1} \lambda_2^{\beta_2 + 1} F(\beta_1, \beta_2, y/\eta). \qquad (3.6.10)$$

F is a function defined by

$$F(\beta_1,\beta_2,z) = \int_0^\infty du_1\, du_2\, u_1^{-\beta_1-1} u_2^{-\beta_2-1}\, e^{i(u_1 u_2 z + u_1 + u_2)}$$

$$= \Gamma(-\beta_1)\Gamma(-\beta_2)\Phi(\beta_1,\beta_2,z), \qquad (3.6.11)$$

where the Γs are gamma functions and Φ is related to the confluent hypergeometric function, $_1F_1$, by

$$\Phi(\beta_1,\beta_2,z) = (-i)^{\beta_1} z^{\beta_1} [\Gamma(\beta_1-\beta_2)/\Gamma(\beta_1+1)]\sin\pi\beta_2$$
$$\times {}_1F_1(-\beta_1,\beta_2-\beta_1+1,-iz^{-1}) + (\beta_1 \leftrightarrow \beta_2).$$
$$(3.6.12)$$

The integral (3.6.10) must have poles at $\beta_1 = \alpha_1(t_1)$, $\beta_2 = \alpha_2(t_2)$, to give the Regge poles of fig. 2.9.2. These poles must arise from divergences at $\lambda_1 = 0$, $\lambda_2 = 0$, respectively, so that the weight function ψ must contain a term behaving like

$$\lambda_1^{-\alpha_1-2}\lambda_2^{-\alpha_2-2}\hat{f}(y;t_1,t_2). \qquad (3.6.13)$$

The two-Reggeon–particle coupling is therefore proportional to

$$\int_0^\infty dy f_{\alpha_1\alpha_2}(y;t_1,t_2)\Phi(\alpha_1,\alpha_2,y/\eta), \qquad (3.6.14)$$

with

$$f_{\alpha_1\alpha_2}(y,t_1,t_2) = \frac{\Gamma(-\alpha_1)\Gamma(-\alpha_2)}{g_1(t_1)g_2(t_2)}\hat{f}(y;t_1,t_2), \qquad (3.6.15)$$

the g-functions being the two-particle–Reggeon couplings which must be factored out. Equation (3.6.14) provides a standard form for the coupling function, with f playing the role of a weight function. Certain general properties follow from (3.6.15). When α_1 or α_2 is a non-negative integer n, the properties of $_1F_1$ make (3.6.15) a polynomial of degree n in η^{-1} or, equivalently, $\cos\omega$. This corresponds to the ω-dependence to be expected from the fact that at integer α the Regge trajectory gives a particle of spin n.

For general values of α_1 and α_2 (3.6.14) takes the form

$$\eta^{-\alpha_1}A_1(\eta) + \eta^{-\alpha_2}A_2(\eta), \qquad (3.6.16)$$

where A_1 and A_2 are entire functions of η depending on α_1 and α_2. It is important to note that (3.6.16) explicitly exhibits the presence of cuts in η in the coupling function $f_{\alpha_1\alpha_2}$.

The decomposition (3.6.16) into the sum of two terms has been explained by Halliday (1971). When multiplied by $s_1^{\alpha_1} s_2^{\alpha_2}$ the two terms of (3.6.16) give functions which at fixed s have cuts only in s_2 and s_1 respectively. Thus the double discontinuity in s_1 and s_2 of (3.6.17) is zero. There is a general theorem, known as the Steinmann relation (Araki, 1960), which requires that double discontinuities in subenergies corresponding to overlapping sets of particles should vanish.

Signature properties

So far we have considered the exchange of unsignatured Regge poles (Collins, 1977). The existence of cuts in η complicates the discussion of signature properties in multi-Reggeon exchange (Drummond, Landshoff & Zakrzewski, 1969b).

In the unsignatured amplitude of fig. 2.9.2 η is to be taken above its cut, drawn along the positive real axis. This can be seen as follows. The cut in η arises from the cuts in s, s_1, s_2. If all these variable were to tend to $-\infty$ they would be free of their cuts. So would η, for it would then be negative. To obtain the physical limit s, s_1, and s_2 must each be continued in the upper half plane from its negative value, that is, $s \to (-s)e^{-i\pi}$, etc. Then $\eta \to (-\eta)e^{-i\pi}$ and is also above its cut.

To evaluate the signatured amplitude one must add to fig. 2.9.2 the three additional diagrams shown in fig. 3.6.2, where τ_i is the signature factor (± 1) of the ith Regge pole. The two crossed variables, $(p' - p_1)^2$ and $(p_3 - p)^2$ associated with the second diagram, which are the analogues of s and s_1, are in fact negative in the physical region. Thus for this diagram only s_2 needs to be continued to positive values above its cut. This takes η

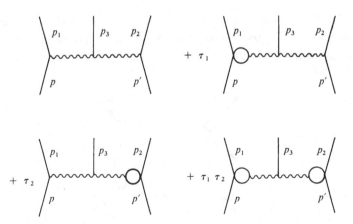

FIGURE 3.6.2 The generalisation of fig. 2.9.2 to the case of signatured Regge pole exchange.

again on top of its cut, as it is also for the third diagram of fig. 3.6.2. However for the fourth diagram the variable η is free from cuts when $(p_1 + p_2)^2$, $(p_3 + p)^2$, $(p_3 + p')^2$ tend to $-\infty$. This corresponds to s_1, $s_2 \to \infty$, $s \to -\infty$. Thus η in the physical region is taken onto its cut by continuing s alone in the upper half plane. By (3.6.5) this brings η *below* its cut. This mismatch in the placing of η in relation to its cut between the fourth diagram of fig. 3.6.2 and the other three diagrams means that the phase of the amplitude is not given by the signature factors alone, in contrast with the case for single-Reggeon exchange in two-particle scattering.

The idea of using a Fourier transform like (3.6.8) to express analytic properties and thus to obtain a spacelike representation for Reggeon couplings, like (3.6.55), is one that finds several applications. For example it can also be used to discuss the coupling of three Reggeons in the triple-Regge limit (Landshoff & Zakrzewski, 1969).

3.7 Regge cuts

In section 2.7 we found that AFS type of Regge cut was not present on the physical sheet but that a more complicated diagram, identified by Mandelstam (1963), does have that property. We will now give a simple explanation, due in this form to Rothe (1967), of why this is so.

Elementary theory

Consider the integral associated with fig. 3.7.1 in which a circulating loop momentum k passes through the momenta q_i. The integral over the four components k_μ may be replaced by an integral over the four invariants

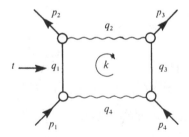

FIGURE 3.7.1 The AFS diagram.

q_i^2 (Drummond, 1963)

$$\int d^4 k \to \int \frac{dq_1^2 dq_2^2 dq_3^2 dq_4^2 \, \theta(\Delta)}{\Delta^{1/2}(q_1^2, q_2^2, q_3^2, q_4^2; s, t)}, \qquad (3.7.1)$$

where the Jacobian $\Delta^{1/2}$ is given by

$$\Delta = \det[q_i \cdot q_j]$$
$$\sim s^2 \lambda(t; q_2^2, q_4^2), \qquad s \to \infty, \qquad (3.7.2)$$

with

$$\lambda(t; t_1, t_2) = t^2 + t_1^2 + t_2^2 - 2tt_1 - 2tt_2 - 2t_1 t_2. \qquad (3.7.3)$$

Equation (3.7.2) comes from noting that since k appears linearly in each q_i the Jacobian is proportional to the 4×4 determinant composed of the four components of the four q_is. Multiplying this determinant by its transpose gives (3.7.2).

Assuming that the limit $s \to \infty$ can be taken under the integral sign, the amplitude associated with fig. 3.7.1 has the asymptotic form

$$\int_{-\infty}^{\infty} \frac{dq_1^2}{q_1^2 - m^2 + i\varepsilon} \int_{-\infty}^{\infty} \frac{dq_3^2}{q_3^2 - m^2 + i\varepsilon} \int \frac{dq_2^2 dq_4^2 \, \theta(\lambda)}{\lambda^{1/2}(t; q_2^2, q_4^2)}$$
$$\times g_1(q_1^2, q_2^2) g_1(q_2^2, q_3^2) g_2(q_3^2, q_4^2) g_2(q_4^2, q_1^2)$$
$$\times s^{\alpha_1(q_2^2) + \alpha_2(q_4^2) - 1}, \qquad (3.7.4)$$

where the g_is are the coupling functions of the Regge trajectory α_i, and for generality we have considered the possibility of two different Reggeons being exchanged. The expression (3.7.4) appears to manifest an asymptotic behaviour proportional to

$$\int \frac{dt \, dt_1 \, dt_2}{\lambda^{1/2}(t; t_1, t_2)} s^{\alpha_1(t_1) + \alpha_2(t_2) - 1}, \qquad (3.7.5)$$

which, because it corresponds to the integral of a variable power behaviour, would correspond to a Regge cut singularity in the complex l-plane (see (2.1.3)). However the whole expression (3.7.4) actually vanishes. To see this consider the q_1^2 integration. The coupling functions g_i are expected to have only right hand cuts in q_1^2 which are approached from above in the physical limit. Thus the q_1^2-integral is as shown in fig. 3.7.2, where the pole is associated with the propagator in (3.7.4) and the cuts come from the gs. The contour can be completed at infinity in the upper half plane to give zero because the rapid decrease of the gs will give the necessary

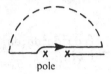

FIGURE 3.7.2 The q_1^2-integration

convergence as $|q_1^2| \to \infty$. Of course the q_3^2-integration in (3.7.4) also gives zero for a similar reason.

Notice that if the two-particle discontinuity were being evaluated rather than the amplitude, the propagators $(q_i^2 - m^2 + i\varepsilon)$, $i = 1, 3$, would be replaced by δ-functions $- 2\pi \delta(q_i^2 - m^2)$ and there would be no vanishing integration. Thus, as stated in section 2.7, the discontinuity has the cut behaviour. However the presence of the Regge cut in this discontinuity does not tell us whether it is present in the amplitude on the physical sheet, or the unphysical sheet through the two-particle cut, or both. In fact we see that the second of these possibilities must be the correct one.

It is clear that a Regge cut can only arise if both q_1^2 and q_3^2 possess also left hand cut singularities (as in fig. 3.7.3) which prevent contour closing. This is achieved in the Mandelstam diagram fig. 3.7.4, where q_1 and q_3 are understood to be the total momentum flowing through each cross respectively. It is clear that the general condition for the presence of a Regge cut on the physical sheet is that the two Reggeons should couple to both the pairs of incident and outgoing particles through subamplitudes with both s- and u-channel singularities. The crosses of fig. 3.7.4 are just the most simple example of such an amplitude.

FIGURE 3.7.3 The form of q_1^2-integration required to give a non-vanishing Regge cut.

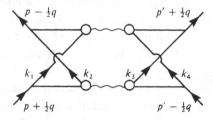

FIGURE 3.7.4 The Mandelstam diagram.

Sudakov parameter discussion

It is instructive to consider fig. 3.7.4 in more detail by using the Sudakov variable techniques first applied to such problems by Gribov (1968). The diagram is to be interpreted in a hybrid sense. The solid lines represent Feynman propagators while the Reggeons and their coupling 'blobs' together indicate subamplitudes which if evaluated in the appropriate high energies limit will exhibit Regge asymptotic behaviour. The momenta are chosen as in the figure and, if for simplicity we consider the case of equal-mass external particles, this implies

$$p \cdot q = p' \cdot q = 0,$$
$$p^2 = p'^2 = m^2 - \tfrac{1}{4}q^2 \equiv \tau. \tag{3.7.6}$$

The vector q is spacelike and the momentum transfer $t = q^2 \leqslant 0$. The internal momenta k_i will be written in the form

$$k_i = x_i p + y_i p' + \kappa_i, \quad \kappa_i \cdot p = \kappa_i \cdot p' = 0. \tag{3.7.7}$$

At high s and fixed t the external momenta almost lie in a two-space, because the energy and longitudinal components are large and the components in the transverse space, in which q and the κ_i lie, are finite. The usefulness of the Sudakov technique largely rests on the simple way in which it distinguishes between these degrees of freedom.

In terms of the Sudakov parameters

$$k_i^2 = x_i y_i s + (x_i - y_i)^2 \tau + \kappa_i^2, \tag{3.7.8}$$

$$d^4 k_i \rightarrow \tfrac{1}{2} s \, dx_i \, dy_i \, d^2 \kappa_i. \tag{3.7.9}$$

Momentum conservation implies that

$$\sum x_i = 1, \quad \sum y_i = 1, \quad \sum \kappa_i = 0. \tag{3.7.10}$$

The squares of the momenta in the lines of the left hand cross are given by

$$\left.\begin{aligned}
k_1^2 &= x_1 y_1 s + (x_1 - y_1)^2 \tau + \kappa_1^2, \\
(p + \tfrac{1}{2}q - k_1)^2 &= (x_1 - 1)y_1 s + (x_1 - y_1 - 1)^2 \tau + (\kappa_1 - \tfrac{1}{2}q)^2, \\
k_2^2 &= x_2 y_2 s + (x_2 - y_2)^2 \tau + \kappa_2^2, \\
(p - \tfrac{1}{2}q - k_2)^2 &= (x_2 - 1)y_2 s + (x_2 - y_2 - 1)^2 \tau + (\tfrac{1}{2}q + \kappa_2)^2.
\end{aligned}\right\} \tag{3.7.11}$$

It is a basic assumption of this approach that the Reggeon coupling functions decrease rapidly as the virtual masses become large. Thus the significant region of integration will be one in which all the terms of (3.7.11) are finite. By inspection we see that this is given by

$$y_1 = \bar{y}_1/s, \quad y_2 = \bar{y}_2/s, \quad x_1, x_2 \text{ finite.} \tag{3.7.12}$$

Similarly, consideration of the momenta in the lines of the right hand cross leads to the conditions

$$x_3 = \bar{x}_3/s, \quad x_4 = \bar{x}_4/s, \quad y_3, y_4 \text{ finite.} \tag{3.7.13}$$

The equations of (3.7.10), (3.7.12) and (3.7.13) together imply that

$$x_1 + x_2 - 1 = O(s^{-1}), \quad y_3 + y_4 - 1 = O(s^{-1}), \tag{3.7.14}$$

and hence the momenta flowing through the Reggeons have finite squares

$$(k_1 + k_2 + p \pm \tfrac{1}{2}q)^2 = (\kappa_1 + \kappa_2 \pm \tfrac{1}{2}q)^2. \tag{3.7.15}$$

In contrast the energies across the Reggeons are large

$$(k_1 + k_4)^2 \sim x_1 y_4 s, \quad (k_2 + k_3)^2 \sim x_2 y_3 s. \tag{3.7.16}$$

In other words the two subamplitudes are indeed evaluated in a Regge limit.

The asymptotic behaviour associated with fig. 3.7.4 is therefore found to be

$$\tfrac{1}{2} \int d^2 \kappa [\hat{X}(\kappa)]^2 s^{\alpha[(\kappa - q/2)^2] + \alpha[(\kappa + q/2)^2] - 1}. \tag{3.7.17}$$

The two $\hat{X}$-factors are associated with the two crosses and take the form

$$\begin{aligned}
\hat{X}(\kappa) = \tfrac{1}{2} \int & d^2 \kappa_1 d^2 \kappa_2 \delta(\kappa_1 + \kappa_2 - \kappa) dx_1 dx_2 d\bar{y}_1 d\bar{y}_2 \\
& \times \delta(x_1 + x_2 - 1) x_1^{\alpha[(\kappa + q/2)^2]} x_2^{\alpha[(\kappa - q/2)^2]} g_1 g_2 \\
& \times [x_1 \bar{y}_1 + x_1^2 \tau + \kappa_1^2 - m^2 + i\varepsilon]^{-1} \\
& \times [-x_2 \bar{y}_1 + x_2^2 \tau + (\kappa_1 - \tfrac{1}{2}q)^2 - m^2 + i\varepsilon]^{-1} \\
& \times [x_2 \bar{y}_2 + x_2^2 \tau + \kappa_2^2 - m^2 + i\varepsilon]^{-1} \\
& \times [-x_1 \bar{y}_2 + x_1^2 \tau + (\kappa_2 + \tfrac{1}{2}q)^2 - m^2 + i\varepsilon]^{-1}. \tag{3.7.18}
\end{aligned}$$

Here g_1 and g_2 are the Reggeon coupling functions and depend on the momenta squared in the appropriate lines. The form (3.7.18) is that naturally associated with the left hand cross. The right hand cross gives $\hat{X}$ of identical form with the integral given by the change of variables

$$x_1, x_2 \to y_3, y_4, \quad \bar{y}_1, \bar{y}_2 \to \bar{x}_3, \bar{x}_4. \tag{3.7.19}$$

Consider the $\bar{y}_1$-integration in (3.7.18). It runs from $-\infty$ to $+\infty$ and there must be singularities on both sides of the contour to avoid its being completed at infinity to give zero (compare the discussion of section 3.2).

The singularities in $\bar{y}_1$ arise from the first two denominators in (3.7.18) (and also corresponding singularities in g_1, g_2 which obey the same $+ \mathrm{i}\varepsilon$ prescription). Thus the integral will only be non-zero if

$$0 < x_1 < 1, \qquad (3.7.20)$$

and we may similarly deduce that from the $\bar{y}_2$-integration that

$$0 < x_2 < 1, \qquad (3.7.21)$$

and for the right hand cross that

$$0 < y_3 < 1, \quad 0 < y_4 < 1. \qquad (3.7.22)$$

Signature can easily be incorporated into the diagram. For boson trajectories the rule is that the signature of the Regge cut is the product of the signatures of the two exchanged Reggeons (Gribov, 1968; Polkinghorne, 1968). For baryon trajectories the rule is modified. Branson (1969) showed that the signature of a two-baryon cut is minus the product of the signatures of the baryons. (Formally this can be neatly represented by assigning signatures $\pm \mathrm{i}$ to baryons!)

The Gribov–Pomeranchuk essential singularity

Mandelstam suggested that the reason for the existence of Regge cuts lies in the need to cure a phenomenon called the GP disease (Gribov & Pomeranchuk, 1962). In section 2.6 we saw that cross-diagrams on iteration lead to multiple poles at $l = -1$. This can be thought of as due to the single cross corresponding to a single pole at $l = -1$ in the wrong signatured (even) amplitude which on iteration through two-particle intermediate states then leads to multiple poles. These overcome the vanishing signature factor and so give contributions to the asymptotic behaviour which are just the pinch terms of section 2.6. The increasing multiplicity of poles at $l = -1$ without factorised residues can then be interpreted as the occurrence of an essential singularity at $l = -1$.

The unitarity equation across the two-particle normal threshold takes the form

$$a^{(1)} - a^{(2)} = \rho a^{(1)} a^{(2)}, \qquad (3.7.23)$$

where ρ is the two-particle phase space factor independent of l, and $a^{(1)}$ and $a^{(2)}$ are the partial wave-amplitudes above and below the normal threshold cut, continued to complex l. If there is no other cut near the normal threshold as l is continued to -1 then the pole at $l = -1$ due to the single cross is present in both $a^{(1)}$ and $a^{(2)}$, since they are just complex conjugates of each other. Then (3.7.23) implies that $a^{(1)}$ or $a^{(2)}$, or both, in fact have a double pole, since the right hand side does; but then repeating the argument leads to a triple pole, etc., and hence to the infinite condensation of poles at $l = -1$ which is the essential singularity. It is clear that the discussion of section 2.6 is just a diagrammatic version of this argument.

The conclusion can be avoided if there is another cut in the positive-signatured amplitude which covers the two-particle normal threshold cut as $l \to -1$, as in fig. 3.7.5. Then $a^{(2)}$ is in the region enclosed by the two cuts. It is no longer the complex conjugate of $a^{(1)}$ so that it is now not possible to argue that it has a pole. The pole is only present in $a^{(1)}$ (and in the complex conjugate amplitude which is on the underside of both cuts). Thus (3.7.23) no longer leads to an essential singularity.

Diagrams like fig. 3.7.4 are simply crosses connected, not by particles (as in fig. 2.6.1), but by Reggeons. Mandelstam (1963) suggested that the cuts so generated would play the shielding role of fig. 3.7.5. This is so, but the way in which it happens is rather complicated. First, just as one must consider iterations of fig. 2.6.1 to generate the GP disease, so it is necessary to consider iterations of fig. 3.7.4, where a chain of crosses is joined by pairs of Regge poles. This is readily done by generalisations of the method given above (Landshoff & Polkinghorne, 1969).

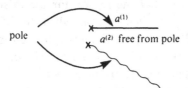

FIGURE 3.7.5 The t-plane near $l = -1$ if the two-particle normal threshold cut is covered by a second cut.

Then it turns out that it is not the Reggeon–Reggeon cut of fig. 3.7.4 which in general plays the shielding role but an associated singularity, the Reggeon–particle cut, analogous to fig. 2.7.2a, with the particle corresponding to a pole on one of the Regge trajectories (Schwarz, 1967). It can then be verified explicitly (Olive & Polkinghorne, 1968) that this cut has just the property necessary to cure the GP disease.

3.8 Reggeon field theory

Fig. 3.7.4 represents just one of the simplest diagrams that can be discussed by using the hybrid techniques. The method can be applied to a wealth of

such diagrams and from this study an algorithm is deduced which is
the basis of *the Gribov Reggeon calculus*. The calculus is expressed in
terms of diagrams like fig. 3.8.1. The wavy internal lines represent
Reggeons. They have a directed sense, identifying the vertex at which a
Reggeon is emitted and the vertex at which it is absorbed. This means
that the two topologically identical diagrams of fig. 3.8.1 are counted as
distinct. The three-Reggeon vertices are represented by non-planar
constructs like fig. 3.8.2, formed from propagators, just as the two-particle
two-Reggeon vertices are represented by the crosses of fig. 3.7.4. The
three-Reggeon vertices of fig. 3.8.2 must be distinguished from the vertices
introduced in the discussion of multi-Regge limits given in section 3.6.
These latter, of which the two-Reggeon one-particle vertex of fig. 3.6.1
is an example, can be represented by simple planar subdiagrams. The
reason for this is that these latter multi-Regge vertices depend on further

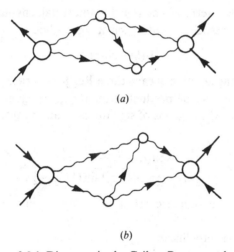

(a)

(b)

FIGURE 3.8.1 Diagrams in the Gribov Reggeon calculus.

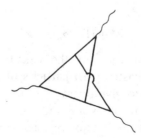

FIGURE 3.8.2 A model for the three-Reggeon vertex.

variables of the Toller η type. These variables correspond to degrees of freedom associated with the helicities of the Reggeons. In the loops of the Reggeon calculus the helicities are implicitly summed over. Under this summation only the more complicated non-planar insertions give a non-vanishing answer (Landshoff, 1970).

It is allowed in the calculus to have four or more Reggeons incident at a vertex, though it is possible to argue that, for some applications, diagrams with only three-Reggeon vertices lead to the dominant effect.

It is only appropriate in this monograph to outline the general features of the calculus. For details reference should be made to existing comprehensive reviews (Abarbanel, Bronzan, Sugar & White, 1975; Baker & Ter-Martirosyan, 1976). Each Reggeon line has associated with it a two-dimensional (transverse) momentum κ_i and an 'energy' $1 - l_i$, where l_i is an angular momentum variable. These variables are conserved at each vertex with the transverse momentum q and 'energy' $1 - l$ being fed in from the external lines (l being the total angular momentum). The propagator associated with each Reggeon line is

$$1/(l_i - \alpha(\kappa_i^2)). \tag{3.8.1}$$

There is a coupling function at each three-Reggeon vertex, $r_{l_1 l_2 l_3}(\kappa_1^2, \kappa_2^2, \kappa_3^2)$, which vanishes unless the product of the Reggeon signatures is $+1$. For each pair of *emitted* Reggeons of signatures τ_1 and τ_2 there is a signature factor

$$\gamma_{l_1 l_2} = \frac{\cos\frac{1}{2}\pi(l_1 + l_2 + 1 - \frac{1}{2}\tau_1 - \frac{1}{2}\tau_2)}{\sin(\frac{1}{2}\pi l_1 - \frac{1}{4}\pi - \frac{1}{4}\pi\tau_1)\sin(\frac{1}{2}\pi l_2 - \frac{1}{4}\pi - \frac{1}{4}\pi\tau_2)}. \tag{3.8.2}$$

'Energy' and momentum circulating in a loop are integrated over in the usual way.

If the trajectories are linear

$$\alpha(\kappa^2) = \alpha_0 + \alpha'\kappa^2, \tag{3.8.3}$$

the 'energy' can be written

$$E(\kappa) = \alpha'\kappa^2 + (1 - \alpha_0), \quad \kappa^2 = -\kappa^2. \tag{3.8.4}$$

This suggests a reformulation of the calculus in which Reggeons correspond to non-relativistic quasi-particles of mass $(2\alpha')^{-1}$ and with an energy gap $(1 - \alpha_0)$, to give quasi-particle energy as in (3.8.4). The vertices r correspond to the emission and absorption of quasi-particles. The particle creation and annihilation processes can be represented in the usual way by the creation and annihilation operators of a quantum field theory set up in the (two-dimensional space + one-dimensional 'time') conjugate to the (transverse momentum + 'energy') variables of the calculus. The diagrams of this field theory are then the Reggeon diagrams of the calculus, but the theory is of rather odd kind because the ϕ^3-vertex corresponding to r turns out to have an imaginary coupling constant, that is the 'interaction' is antihermitean.

If $\alpha_0 = 1$ (as it will for diffractive scattering with conventional Pomeron exchange) the energy gap vanishes. The consequent infra-red singularities control the behaviour in the forward direction where zero transverse momentum is exchanged. The main thrust of the theory is to master the subtleties of this situation and so to understand the nature of diffractive scattering at ultra-high energy. However it is not in fact necessary for the 'bare' Pomerons exchanged to have $\alpha_0 = 1$. They can also have $\alpha_0 > 1$, for the field theory is believed to respect unitarity so that the 'renormalised' singularity is brought back to $l = 1$ in compliance with the Froissart bound. For an account of these highly complicated matters reference should be made to the reviews by Abarbanel *et al.* (1975) and Baker & Ter-Martirosyan (1976).

4
Discontinuities

4.1 Introduction

We are often concerned to calculate the discontinuity of an amplitude in a given channel in order to use it to calculate a cross-section via Mueller's (1970) generalisation of the optical theorem. In the preceding chapters our technique normally has been to calculate the amplitude itself and then to take an appropriate imaginary part. The first method given in section 3.2 for calculating the contribution of the handbag diagram to deep inelastic scattering is typical of this procedure. However it is also possible to calculate the discontinuity directly. The second method of section 3.2 for treating the handbag diagram is an example of this. In this chapter we present a more systematic account of discontinuity calculations. In the next section we shall establish a remarkable theorem which states that in the calculation of the differential cross-section the final state corrections due to Pomeron exchange actually cancel for hard scattering processes.

4.2 Final state interactions

In the deep inelastic processes of sections 3.2 and 3.4 it is an important question whether the cross-section is modified by taking into account the possibility of final state interactions between the scattered constituents. In section 3.2 we raised the problem of whether the exchange of a Pomeron between the struck quark and the residual 'core' of the hadron would give an additional contribution to the scaling structure functions, but we deferred the detailed answer till this section. Such a Pomeron exchange would represent a diffractive final state interaction between the struck quark and the hadronic residue. Similarly in the parton hard scattering processes of section 3.4 we can ask whether final state interactions between pairs of outgoing systems give extra contributions to the differential cross-section. A simple power-counting argument again shows that such interactions could only contribute at the leading order (3.4.16) if they corresponded to Pomeron exchange.

In our present state of knowledge these questions can only receive interim answers. This is due to our ignorance of the mechanism which produces *quark confinement*. For example, in deep inelastic electroproduction the struck quark does not materialise as an observable quark but the final state is composed of conventional hadrons. Some triality-sorting final state interaction must be responsible for this. Various models have been investigated (Preparata, 1973; Polkinghorne 1975; Jaffe & Patrascioiu, 1975) which encourage the view that the triality-sorting confining forces do not alter the cross-sections from their values calculated from simple quark parton models. However there is no description available in fundamental terms for the effects of the confinement, so that its not changing the results of chapter 3 remains a conjecture.

In this section we must be content to consider final state interactions of a conventional kind, that is, diffractive effects mediated by the exchange of a Pomeron, regarded as a Regge pole whose coupling function to quarks (as well as to hadrons) is assumed to decrease rapidly when the particles are far off shell.

Diffractive corrections to deep inelastic scattering

We now consider whether the exchange of a Pomeron, as in fig. 4.2.1, gives a contribution to the cross-section in the Bjorken limit comparable with that of the handbag diagram.

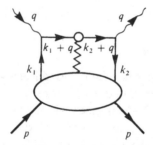

FIGURE 4.2.1 A diffractive correction to the handbag diagram. The zig-zag line represents a Pomeron.

The transformations (3.2.9), (3.2.12), (3.2.28) applied to k_1 and k_2 make all the variables

$$\left.\begin{array}{l} (p-k_1)^2, k_1^2, (k_1+q)^2, \\ (p-k_2)^2, k_2^2, (k_2+q)^2, \end{array}\right\} \tag{4.2.1}$$

finite. This costs v^{-2} but the factor of v from the Pomeron gives a net

ν^{-1}, comparable with the handbag. However we shall see that the integral multiplying this vanishes.

The variables $\bar{x}_1, \bar{x}_2$ appear only in $(k_1 + q)^2, (k_2 + q)^2$, respectively, which are the masses of the quarks coupling to the Pomeron. The discontinuity arises from the insertion of a set of intermediate states into the diagram fig. 4.2.1. This can be done in a number of ways; either one or other of the quarks can be involved (in which case its propagator is replaced by an integral of the spectral function ρ over a limited range of $\bar{x}_i$, just as for the handbag) or the intermediate state arises from within the Pomeron itself (compare section 3.5). Diagrammatically these possibilities can be represented by a bar slicing through the diagram (cf. fig. 3.2.7) in three different ways (through the Pomeron or to the left or right of it), the cut lines being those which generate the final state particles. The total discontinuity is the sum of these terms which represent the different ways in which the final state can be built up. Whichever slice is taken there is always at least one of the $\bar{x}_i$-integrations (corresponding to the uncut quark line) which runs unrestrictedly from $-\infty$ to $+\infty$, passing above the singularities in $(k_i + q)^2$. This contour can then be completed at infinity to give zero. (This argument fails if the Pomeron is replaced by a *point*-coupled vector meson since there is then insufficient convergence at infinity to complete the contour.)

Thus no diffractive corrections are present in deep inelastic processes. This result generalises to more complicated situations, as we shall now investigate.

The cancellation theorem

As an example of the cancellation of diffractive effects in the cross-section for a more complicated parton process we consider the diffractive corrections to the Drell–Yan mechanism. The relevant diagram is fig. 4.2.2. In contrast to the case of deep inelastic scattering the individual slices contributing to the discontinuity of fig. 4.2.2 are not separately zero because it is found that there are singularities so disposed as to prevent contour closing. However when all the contributions are added together they give a vanishing correction. This was first shown by Cardy & Winbow (1974) using the AGK picture described in section 4.3. Here we follow a treatment due to DeTar, Ellis & Landshoff (1975). It makes essential use to the ideas of Mueller (1970; see also Polkinghorne, 1972) which avoids the need to consider separate slices.

The parton momenta of fig. 4.2.2 are parameterised in analogy with

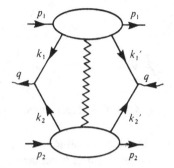

FIGURE 4.2.2 A diffractive correction to the Drell–Yan process.

(3.2.62) (3.2.63), as

$$\left.\begin{aligned} k_i &= x_i p + (y_i/s)p' + \kappa_i, \\ k_i' &= (x_i'/s)p + y_i'p' + \kappa_i', \end{aligned}\right\} i = 1, 2. \qquad (4.2.2)$$

They are subject to the constraint

$$k_1 + k_1' = q = k_2 + k_2', \qquad (4.2.3)$$

so that

$$x_1 = x_2 + O(s^{-1}), \quad y_1' = y_2' + O(s^{-1}). \qquad (4.2.4)$$

The kinematic requirements

$$q^2 > 0, \quad 0 < q_0 < (p + p')_0, \qquad (4.2.5)$$

lead to the expected conditions

$$0 < x_1 < 1, \quad 0 < y_1' < 1. \qquad (4.2.6)$$

The parametrisation (4.2.2) makes the variables

$$\left.\begin{aligned} \mu_i^2 &\equiv k_i^2 = x_i y_i + x_i^2 m^2 + \kappa_i^2, \\ s_i &\equiv (p - k_i)^2 = (x_i - 1)y_i + (x_i - 1)^2 m^2 + \kappa_i^2, \\ u_i &\equiv (p + k_i - k_j)^2 = y_i - y_j + m^2 + (\kappa_i - \kappa_j)^2; \end{aligned}\right\} \begin{aligned} i &= 1, 2; \\ j &= 2, 1; \end{aligned} \qquad (4.2.7)$$

finite, together with the corresponding primed variables, $\mu_i'^2 \equiv k_i'^2$, etc., with $x \to y'$, $y \to x'$. Simple power counting again indicates an expression comparable with the Drell–Yan cross-section (3.2.65). However it is necessary to investigate whether or not the coefficient is non-zero.

According to Mueller (1970) the total contribution of fig. 4.2.2 to the cross-section is given by the discontinuity in the missing mass variable $(p + p' - q)^2$ of an unphysical amplitude associated with the figure. The exact discontinuity needed is specified by fig. 4.2.3. The $\pm$ indicate

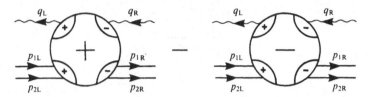

FIGURE 4.2.3 The Mueller discontinuity.

the $\pm i\varepsilon$ prescriptions for the variables $(p + p')^2$, q^2 and (centrally) $(p + p' - q)^2$. For the subenergies it is necessary to discriminate between the (numerically equal) momenta on the left (p_L, p'_L, q_L) and on the right (p_R, p'_R, q_R), since the associated cuts have to be treated differently. The different prescriptions for $(p + p' - q)^2$ in the two terms of fig. 4.2.3 evaluate the Mueller discontinuity.

If the discontinuity is in fact zero we expect to show this by a contour closing argument. This means that we have to find the $i\varepsilon$ prescriptions for the internal variables (4.2.7) which are implied by the prescriptions indicated for the external variables in fig. 4.2.3. The two sets of prescriptions are linked together because the singularities in the external variables are generated by integrating over the singularities in the internal variables (Eden *et al.*, 1966, ch. 4). There is a general theory which establishes how such connections are to be made (Bloxham, Olive & Polkinghorne, 1969). For our present purpose it will be sufficient to summarise the results of the general analysis as it applies to the integral of fig. 4.2.2.

A first step is to identify which internal singularities generate which external singularities. A simple rule is based on drawing a Feynman-like diagram whose 'propagators' carry the momenta associated with internal invariants of interest. We are currently concerned only with normal threshold singularities. If the Feynman-like diagram can be contracted (by short circuiting some lines; that is, in the language of chapter 1, putting their αs equal to zero) to form a self-energy part in an external variable, then the normal threshold in that external variable can be

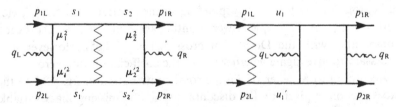

FIGURE 4.2.4 Feynman-like diagrams to indicate how internal singularities generate external singularities.

generated by the singularities in the internal invariants associated with the non-contracted lines which survive in the self-energy part. Feynman-like diagrams relevant to fig. 4.2.2 are given in fig. 4.2.4. From studying their contractions we see that the following pairs of singularities in internal variables will generate singularities in the indicated external variables:

$$\left.\begin{aligned}
\mu_1^2, \mu_1'^2 &\to q_L^2, \\
\mu_2^2, \mu_2'^2 &\to q_R^2, \\
s_1, s_1' &\to M^2 \equiv (p + p' - q)^2, \\
s_2, s_2' &\to M^2 \equiv (p + p' - q)^2, \\
u_1, u_1' &\to s_L \equiv (p_L + p_L')^2, \\
u_2, u_2' &\to s_R \equiv (p_R + p_R')^2.
\end{aligned}\right\} \quad (4.2.8)$$

Having identified these connections between internal and external variables, we make use of the fact that a $+i\varepsilon$ ($-i\varepsilon$) prescription for a normal threshold in an external variable requires matching $+i\varepsilon(-i\varepsilon)$ prescriptions to be assigned to both the internal variables associated with it. Thus the first term of fig. 4.2.3 will imply the following prescriptions for internal variables:

$$\left.\begin{aligned}
+i\varepsilon &: \mu_1^2, \mu_1'^2, s_1, s_1', u_1, u_1', s_2, s_2'; \\
-i\varepsilon &: \mu_2^2, \mu_2'^2, u_2, u_2'.
\end{aligned}\right\} \quad (4.2.9)$$

The variable y_2 appears with a positive coefficient in μ_2^2, u_2 and with a negative coefficient in s_2, u_1. (4.2.9) implies therefore that y_2 is disposed with a $-i\varepsilon$ prescription with respect to all its singularities. Therefore the y_2-integral can be completed in the lower half plane to give zero. The x_2'-integral also gives zero in the first term of fig. 4.2.4. The second term of the figure is zero also since both the y_1- and x_1'-integrations are found to vanish. Thus the Mueller discontinuity is zero and no net diffractive correction results from fig. 4.2.2.

Instead of using the Mueller prescription we could have evaluated the contributions that correspond to slicing the diagram through the Pomeron or to the right or left of it as we did for fig. 4.2.1. These are the contributions associated by unitarity with the different sets of intermediate state insertions corresponding to the slices. Since the integral of the unitarity integral is of the form $T_i \cdot T_f^\dagger$, with T a physical amplitude, the internal $i\varepsilon$ prescriptions are, in this case, $+i\varepsilon$ to the left of the slice (T_i) and $-i\varepsilon$ to the right ($T_f^\dagger$). It is then straightforward to verify that each slice gives a non-vanishing contribution. For example if the slice passes through the Pomeron or to the left of it, the variable s_2 acquires a $-i\varepsilon$ prescription, in contrast to (4.2.9). This prevents the y_2-integration vanishing in this case. If the slice passes to the right of the Pomeron the y_2-integration is of restricted range and so again does not vanish. In a similar way one can verify that all other integrations for each slice give non-vanishing contributions. The Mueller argument implies that these non-vanishing unitarity contributions must cancel when added together. (The two Ts are in general different, so that $T_i \cdot T_f^\dagger$ is not positive definite, except for the slice passing through the Pomeron.) This cancellation can be verified explicitly (DeTar, Ellis & Landshoff, 1975).

We have noted that the different slices will correspond to different structures in the final state. These different final states are therefore influenced by the diffractive interactions but in such a way that the total effect cancels in calculating the cross-section.

The arguments given above can readily be adapted to show the absence of diffractive corrections to the hard scattering processes of section 3.4.

4.3 The AGK cutting rules

Abramovskii, Gribov & Kancheli (1974) have discussed the structure of multi-particle states associated with Regge cuts. A very interesting pattern is revealed of which we shall give a summary, basing the discussion for purposes of illustration on the two-Reggeon cut from the Mandelstam diagram fig. 2.7.2. A detailed discussion of this case was given independently by Halliday & Sachrajda (1973). We shall suppose that the exchanged Reggeons are in fact Pomerons.

The Mandelstam diagram will make many distinct contributions to the final state structure corresponding to different normal threshold discontinuities with their respective sets of particles on the mass-shell. These contributions each correspond to different ways of slicing the diagram into two connected parts carrying s-channel external momenta. The cut lines correspond to the intermediate states inserted into the appropriate unitarity integral or, equivalently, to the final state particles of the scattering process.

One such cut is C_1 of fig. 4.3.1. Clearly it corresponds to a finite number (four) of final state particles. Another such cut is C_2 which slices through a Pomeron. Such a cut gives a large number of final state particles, with an average multiplicity,

$$\langle n \rangle_1 \sim a \ln s, \tag{4.3.1}$$

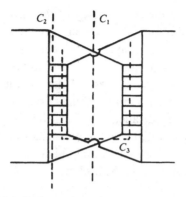

FIGURE 4.3.1 Various slices of the Mandelstam diagram.

as s increases (cf. the discussion following (2.1.21)). Another possibility is cut C_3 which slices through two Pomerons. It will give an even larger final state multiplicity,

$$\langle n \rangle_2 \sim 2a \ln s. \qquad (4.3.2)$$

An important observation is that in the large s limit the most significant contributions come either from slicing right through a Reggeon or from not cutting it at all. A slice like fig. 4.3.2 which exits a ladder in the middle will be subdominant. This is because the large energy necessary to sustain the multi-particle system corresponding to the cut rungs must be carried by a *single* propagator whose mass must therefore be large, $q^2 \sim O(s)$, while if the ladder were cut wholly through, the necessary large energy

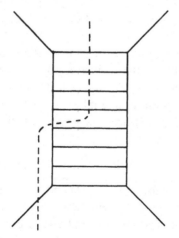

FIGURE 4.3.2 A cut which is not significant in the high energy limit.

would be provided by the total energy of the two finite mass particles entering the top and bottom of the ladder. This large internal momentum suppresses the amplitude. (In hybrid diagrams the propagators are supposed to carry finite mass squared only; see section 3.7.) This argument supposes that the number of cut lines will increase with s. It does not make it clear why a small fixed number of lines could not be cut at the top of fig. 4.3.2. However, detailed investigation of perturbation theory (Halliday & Sachrajda, 1973) shows that such a slice is suppressed below the leading logarithmic approximation. Other approaches to the generation of Reggeons (Degrand, 1975; McLerran & Weis, 1975) all lead to a similar conclusion.

It is found (Abramovskii et al., 1974) that the cuts C_1, C_2, C_3 all give contributions which are multiples of each other, the relative magnitudes being determined by the rule that slicing through a Pomeron multiplies the contribution in which it is uncut by -2. There are two different slices of type C_1 (corresponding to the choice of the pair of lines of the upper cross which are cut; to give an s-channel discontinuity it is then necessary to cut a specific pair in the lower cross); four different slices of type C_2; and only one of type C_3. Thus if $\mathbb{P}_2$ is the contribution due to a C_1 cut, which must be positive since this gives an integrand of type $|T|^2$ in the unitarity integral, the total contribution is

$$2 \cdot \mathbb{P}_2 + 4(-2)\mathbb{P}_2 + 1 \cdot (-2)^2 \mathbb{P}_2 = -2\mathbb{P}_2, \qquad (4.3.3)$$

that is, the sign of the C_1 contribution is just reversed in the complete expression. The negative sign in (4.3.3) corresponds to the physical notion that the double Pomeron cut is a shadowing type of rescattering correction which therefore reduces the cross-section below the value given by single Pomeron exchange.

An interesting consequence of the AGK cutting rules is that the Regge cut does not contribute to the inclusive cross-section in the pionisation region, which is where the large multiplicity originates (Collins, 1977). In principle such contributions might be expected to come from C_2 and C_3 but when (4.3.1) and (4.3.2) are combined with the AGK weights the resulting multiplicity

$$\langle n \rangle \sim 4.(-2)a \ln s + 1.(-2)^2 2a \ln s = 0, \qquad (4.3.4)$$

showing that the total effect must cancel and so give no net contributions to the pionisation plateau. AGK emphasise that this cancellation is a property of the inclusive cross-section; cut effects can be discerned in the individual partial cross-sections but on addition to give the inclusive cross-section these effects cancel out. This is similar to the cancellation effects studied in section 4.2 for hard scattering processes, which again were only present in the cross-section itself.

4.4 Feynman integrals

The discontinuities of Feynman integrals can be evaluated by a prescription first stated by Cutkosky (1960). For a full discussion reference should be made to the theory of Landau singularities (see Eden *et al.*, 1966, ch. 2). For our present purposes it will be sufficient to consider only normal threshold singularities in a given variable. An n-particle normal threshold in the variable s' will be associated with the possibility of contracting the diagram till it takes the form of fig. 4.4.1, that is, an n-lined self-energy part through which the momentum associated with s' flows. By contraction we mean short-circuiting all lines of the diagram except those which carry the q_i, (in the way, for example, that fig. 2.3.2 is obtained from fig. 2.3.1). Then the discontinuity associated with this n-particle normal threshold is obtained from the original Feynman integral by replacing the propagators carrying the q_i by

$$- 2\pi i \delta^+ (q_i^2 - m^2) \equiv - 2\pi i \delta(q_i^2 - m^2)\theta(q_{i0}), \qquad (4.4.1)$$

and leaving all other propagators unchanged, that is, still carrying a $+ \mathrm{i}\varepsilon$ prescription.

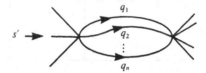

FIGURE 4.4.1 A contraction giving an n-particle normal threshold in s'.

Unitarity formulae

It is clear that the Cutkosky prescription is closely connected with unitarity, as the mass-shell δ-function of (4.4.1) makes evident. Yet unitarity involves the product $TT^\dagger$ of relevant amplitudes and the $T^\dagger$ part must contain propagators with the complex conjugate $- \mathrm{i}\varepsilon$ prescription. However, the rule following (4.4.1) leaves all propagators with $+ \mathrm{i}\varepsilon$. It is instructive to sketch how the unitarity result comes about.

It is due to the fact that diagrams may contain within themselves many different ways in which a normal threshold can be generated. For example, consider the ladder of fig. 4.4.2. This can be contracted to give a two-particle normal threshold in s in many ways, which correspond to any pair of opposite lines along the sides of the ladder being the two lines associated with the contracted self-energy part. In evaluating the discontinuity, the presence of these different ways of generating the singularity, and their interaction with each other, must be taken into account. The rule stated above requires a generalisation for this case which has been worked out (Bloxham *et al.*, 1969). It states that the discontinuity is given by the sum of (a) all single discontinuities evaluated by applying (4.4.1) to a relevant pair of lines; (b) *minus* all two-fold discontinuities evaluated by applying (4.4.1) to *two* pairs of lines; (c) *plus* all three-fold discontinuities evaluated by applying (4.4.1) to *three* pairs of lines; etc. The signs alternate in the sets of terms and all lines not on the mass-shell carry a $+i\varepsilon$ prescription. It is not difficult to see that this rule can be replaced by a prescription which takes only a sum of single discontinuities associated with single pairs of lines but with a $+i\varepsilon$ on propagators to the left of the mass-shell pairs and $-i\varepsilon$ on propagators to the right, (or vice versa). The way this works is illustrated in fig. 4.4.3 where a stroke on a line indicates the mass-shell δ-functions (4.4.1) and $\pm$ indicates $\pm i\varepsilon$ prescriptions. We use the fact that the difference between a $+i\varepsilon$ and $-i\varepsilon$ prescription propagator is just the δ-function (4.4.1).

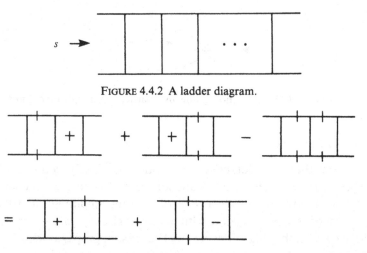

FIGURE 4.4.2 A ladder diagram.

FIGURE 4.4.3 Normal threshold discontinuities summing to give a unitarity-like form.

Discontinuity integrals

It is convenient to use the complex exponential representation (see (1.2.33)) which is well adapted for physical momenta:

$$\frac{i}{q^2 - m^2 \pm i\varepsilon} = \int_0^{\pm \infty} d\alpha \, e^{i\alpha(q^2 - m^2 \pm i\varepsilon)}. \tag{4.4.2}$$

The range of integration in (4.4.2) is determined by the need to use the term $\mp \alpha\varepsilon$ in the exponent as a convergence factor. Equation (4.4.2) makes it clear that $+ i\varepsilon \, (- i\varepsilon)$ prescriptions are associated with positive (negative) values of the αs, respectively. The corresponding representation for (4.4.1) is

$$-2\pi i \delta(q^2 - m^2)\theta(q_0) = -i \int_{-\infty}^{\infty} d\alpha \, e^{i\alpha(q^2 - m^2)}$$

$$\times \frac{1}{2\pi i} \int_{-\infty}^{\infty} \frac{dx}{x - i\varepsilon} e^{2ixq_0}. \tag{4.4.3}$$

Thus mass-shell δ-functions correspond to αs of both signs, with no end point at $\alpha = 0$. This observation is often helpful in indicating the sort of asymptotic behaviour to be expected from discontinuities. The absence of the $\alpha = 0$ end point means that pinches are particularly important.

The new feature of (4.4.3) is the appearances of the second auxiliary variable x, conjugate to the (not manifestly covariant) q_0. The way to treat this additional complication has been shown by Halliday (1978).

It is convenient to introduce a more covariant-appearing form by writing

$$xq_0 = x \cdot q = \alpha r \cdot q, \tag{4.4.4}$$

where x is now a vector,

$$x = (x, 0, 0, 0),$$

and

$$r = (x/\alpha, 0, 0, 0). \tag{4.4.5}$$

The vector r can be thought of as an external vector fed into and out of the line carrying q in the direction of positive energy flow. It resembles the dummy variable a of section 1.3. However we must remember that the quadratic term;

$$\alpha r^2, \tag{4.4.6}$$

which would be necessary to turn αq^2 into $\alpha(q + r)^2$, is missing from the exponent of (4.4.3).

If the r_i are considered as supplementing the set of external momenta p_j, an integral containing terms (4.4.3) can be treated by the symmetric integration techniques of chapter 1. The answer is proportional to

$$\prod_{i,j,k} \int_{-\infty}^{\infty} d\alpha_i \int_0^{\infty} d\alpha_j \int_0^{-\infty} d\alpha_k \int \frac{dx_i}{x_i - i\varepsilon} C^{-2} e^{i\hat{D}/C}. \qquad (4.4.7)$$

Here the α_i, x_i refer to mass-shell lines in accordance with (4.4.3); the $\alpha_j(\alpha_k)$ refer to lines with $+ i\varepsilon$ ($- i\varepsilon$) propagators, in accordance with (4.4.2). C is the usual Feynman function of the αs, and $\hat{D}$ is given by

$$\hat{D} = D(\alpha_1 p) + \sum x_i Y_i + \sum x_i x_i' V_{ij'}. \qquad (4.4.8)$$

D is the Feynman function, and the Y_i are the momentum vectors of section 1.2. The quadratic term of (4.4.8), specified by the matrix V, takes account of the absence of the terms (4.4.6). It is not necessary to know V in detail. For our present purpose it is sufficient to note that it is a known function of the αs and independent of the external momenta p_j.

(There is a technical point to be noted about interpretation of (4.4.7) owing to the fact that having αs of both signs means that C can vanish in the interior of the region of integration. It is necessary to determine how the singularities of C^{-2} are to be avoided by the integration contour. Section 1.2 tells us that C is the determinant of the quadratic form in the symmetric loop momenta. It is given by the product of the eigenvalues of the matrix A. Thus when C vanishes, one of these eigenvalues, λ, must vanish. The formula

$$\int d^4 k e^{i\lambda k^2} = -i\left(\frac{\pi}{\lambda + i\varepsilon}\right)^{1/2}\left(\frac{\pi}{\lambda - i\varepsilon}\right)^{3/2}, \qquad (4.4.9)$$

for a symmetric loop momentum k which is a Lorentz four-vector, then gives the $i\varepsilon$ prescription by which the singularity at $\lambda = 0$ (equivalently, $C = 0$) is to be avoided. The integral (4.4.7) is therefore understood to be defined in this sense.)

The expression given by (4.4.7), (4.4.8) appears to have a Lorentz-frame dependence because of the non-manifestly covariant definition of the vectors x in (4.4.5). However, we know that in fact the answer has to be covariant and the frame-dependence is therefore illusory. The idea of Halliday (1978) is to make use of this apparent frame-dependence by choosing to evaluate (4.4.7) in a particularly simple coordinate system.

For high energy regimes the suitable choice is one in which the momenta Y_i are all large. Since the $V_{ii'}$ are then finite this means that the quadratic terms of (4.4.8) can be neglected. The x_i-integrals can be performed trivially to give

$$\prod_{i,j,k} \int_{-\infty}^{\infty} d\alpha_i \int_0^{\infty} d\alpha_j \int_0^{-\infty} d\alpha_k \, C^{-2} e^{iD/C} \prod_i \theta(Y_{i0}). \qquad (4.4.10)$$

Thus the θ-functions of (4.4.1) simply turn into θ-functions of the large momenta Y_i in this frame of reference.

Halliday (1978) gives interesting examples of the use of (4.4.10).

References

Abarbanel, H. D. I., Bronzan, J. D., Sugar, R. L. & White, A. R., (1975). *Phys. Rep.* **21C**, 121.
Abarbanel, H. D. I. & Itzykson, C., (1969). *Phys. Rev. Lett.* **23**, 53.
Abramovskii, V. A., Gribov, V. N. & Kancheli, O. V., (1974). *Soviet J. Nucl. Phys.* **18**, 308.
Adler, S., (1966). *Phys. Rev.* **143**, 1144.
Altarelli, G. & Rubinstein, H., (1969). *Phys. Rev.* **187**, 2111.
Amati, D., Fubini, S. & Stanghellini, A., (1962). *Nuovo Cim.* **26**, 896.
Araki, H., (1960). *J. Math. Phys.* **2**, 163.

Baker, M. & Ter-Martirosyan, K. A., (1976). *Phys. Rep.* **28C**, 1.
Berman, S. M., Bjorken, J. D. & Kogut, J. B., (1971). *Phys. Rev.* **D4**, 3388.
Bethe, H. A. & Salpeter, E. E., (1951). *Phys. Rev.* **84**, 1232.
Bjorken, J. D., (1966). *Phys. Rev.* **148**, 1467.
Bjorken, J. D., (1969). *Phys. Rev.* **179**, 1547.
Bjorken, J. D. & Drell, S. D., (1965). *Relativistic Quantum Fields*, McGraw-Hill, New York.
Bjorken, J. D. & Wu, T. T., (1963). *Phys. Rev.* **130**, 2566.
Bloxham, M. J. W., Olive, D. I. & Polkinghorne, J. C., (1969). *J. Math. Phys.* **10**, 494.
Bollini, C. G. & Giambiagi, J. J., (1972). *Phys. Lett.* **40B**, 566.
Brandt, R. A. & Preparata, G., (1971). *Nucl. Phys.* **B27**, 541.
Branson, D. B., (1969). *Phys. Rev.* **179**, 1608.
Brodsky, S. & Farrar, G., (1975). *Phys. Rev.* **D11**, 1309.

Cabibbo, N., Parisi, G. & Testa, M., (1970). *Nuovo Cim. Lett.* **4**, 35.
Cardy, J. L., (1970). *Nucl. Phys.* **B17**, 493.
Cardy, J. L., (1971). *Nucl. Phys.* **B33**, 139.
Cardy, J. L. & Winbow, G. A., (1974). *Phys. Lett.* **52B**, 95.
Chang, S.-J. & Fishbane, P. M., (1970). *Phys. Rev.* **D2**, 1084.
Cheng, H. & Lo, C. Y., (1977). *Phys. Rev.* **D15**, 2959.
Cheng, H. & Wu, T. T., (1965). *Phys. Rev.* **140B**, 465.
Chisholm, J. S. R., (1952). *Proc. Camb. Phil. Soc.* **48**, 300.
Collins, P. D. B., (1977). *An Introduction to Regge Theory and High Energy Physics*, Cambridge University Press.
Contogouris, A. P., (1965). *Nuovo Cim.* **36**, 250.
Cornwall, J. M. & Jackiw, R., (1971). *Phys. Rev.* **D4**, 367.
Cutkosky, R. E., (1960). *J. Math. Phys.* **1**, 429.

Degrand, T. A., (1975). *Phys. Rev.* **D11**, 2233.
DeTar, C. E., Ellis, S. D. & Landshoff P. V., (1975). *Nucl. Phys.* **B87**, 176.

Dokshitser, Yu. L., D'Yakonov, D. I. & Troyan, S. I., (1978). *Phys. Lett.* **78B**, 290.

Drell, S. D., Levy, D. J. & Yan T.-M., (1970). *Phys. Rev.* **D1**, 1035, 1617, 2402.

Drell, S. D. & Yan, T.-M., (1970*a*). *Phys. Rev. Lett.* **24**, 181.

Drell, S. D. & Yan, T.-M., (1970*b*). *Phys. Rev. Lett.* **24**, 855.

Drell, S. D. & Yan, T.-M., (1970*c*). *Phys. Rev. Lett.* **25**, 316.

Drummond, I. T., (1963). *Nuovo Cim.* **29**, 720.

Drummond, I. T., (1968). *Phys. Rev.* **176**, 2003.

Drummond, I. T., Landshoff, P. V. & Zakrzewski, W. J., (1969*a*). *Nucl, Phys.* **B11**, 383.

Drummond, I. T., Landshoff, P. V. & Zakrzewski, W. J., (1969*b*). *Phys. Lett.* **28B**, 676.

Eden, R. J., Landshoff, P. V., Olive, D. I. & Polkinghorne, J. C., (1966). *The Analytic S-Matrix*, Cambridge University Press.

Federbush, P. G. & Grisaru, M. T., (1963). *Ann. Phys.* **22**, 263, 299.

Feynman, R. P., (1972). *Photon–Hadron Interactions*, W. A. Benjamin, Inc., New York.

Frenkel, J., Shailer, M. J. & Taylor, J. C., (1979). *Nucl. Phys.* **B148**, 228.

Fritzsch, H. & Gell-Mann, M., (1971). *Proceedings of the International Conference on Duality and Symmetry in Hadron Physics* (ed. E. Gotsmann), Weizmann Science Press, Israel.

Gaisser, T. & Polkinghorne, J. C., (1971). *Nuovo Cim.* **1A**, 501.

Gatto, R., Menotti, P. & Vendramin I., (1972). *Nuovo Cim Lett.* **4**, 79.

Gell-Mann, M., Goldberger, M. L., Low, F. E., Marx, E. & Zachariasen, F., (1964). *Phys. Rev.* **133B**, 145.

Greenman, J. V., (1965). *J. Math. Phys.* **6**, 660.

Gribov, V. N., (1968). *Soviet Phys. J. exp. theor. Phys.* **26**, 414.

Gribov, V. N. & Lipatov, L. N., (1972). *Soviet J. Nucl. Phys.* **15**, 438.

Gribov, V. N. & Pomeranchuk, I. Ya., (1962). *Phys. Lett.* **2**, 239.

Grisaru, M. T., (1977). *Phys. Rev.* **D16**, 1962.

Halliday, I. G., (1963). *Nuovo Cim.* **30**, 177.

Halliday, I. G., (1971). *Nucl. Phys.* **B33**, 285.

Halliday, I. G., (1978). *Nucl. Phys.* **B131**, 477.

Halliday, I. G., Huskins, J. & Sachrajda, C. T., (1974). *Nucl. Phys.* **B83**, 189.

Halliday, I. G. & Sachrajda, C. T., (1973). *Phys. Rev.* **D8**, 3598.

Halliday, I. G. & Saunders, L. M., (1969). *Nuovo Cim.* **60A**, 115.

Hamprecht, B., (1965). *Nuovo Cim.* **40A**, 542.

Hasslacher, D. & Sinclair, D. K. (1971). *Phys. Rev.* **D3**, 1770.

Hinchcliffe, I. & Llewellyn Smith, C. H., (1977). *Phys. Lett.* **66B**, 281.

Islam, J. N., Landshoff, P. V. & Taylor, J. C., (1963). *Phys. Rev.* **130**, 2560.

Jacob. M. & Landshoff, P. V., (1978). *Phys. Rep.* **48**, 285.

Jaffe, R. & Patrascioiu, A., (1975). *Phys. Rev.* **D12**, 1314.

Jauch, J. M. & Rohrlich, F., (1955). *Theory of Photons and Electrons*, Addison-Wesley, Cambridge, U.S.A.

Johnson, K. & Low, F. E., (1966). *Prog. theor. Phys. Suppl.* **37–38**, 74.

Kaschlun, F. & Zoellner, W. (1965). *Nuovo Cim.* **34**, 1618.
Kibble. T. W. B., (1963). *Phys, Rev.* **131**, 2282.
Kogut, J., (1976). *Phys. Lett.* **65B**, 377.

Landshoff, P. V., (1970). *Nucl. Phys.* **B15**, 284.
Landshoff, P. V., (1974). *Phys. Rev.* **D10**, 1054.
Landshoff, P. V. & Polkinghorne. J. C., (1969). *Phys. Rev.* **181**, 1989.
Landshoff, P. V. & Polkinghorne, J. C., (1971*a*). *Nucl. Phys.* **B28**, 240.
Landshoff, P. V. & Polkinghorne, J. C., (1971*b*). *Nucl. Phys.* **B33**, 211.
   (Erratum. *Nucl. Phys.* **B36**, 642.)
Landshoff. P. V. & Polkinghorne, J. C., (1973*a*). *Nucl. Phys.* **B53**, 473.
Landshoff. P. V. & Polkinghorne, J. C., (1973*b*). *Phys. Rev.* **D8**, 927, 4157.
Landshoff. P. V. & Polkinghorne, J. C., (1973*c*). *Phys. Lett.* **44B**, 293.
Landshoff. P. V. & Polkinghorne, J. C., (1974). *Phys. Rev.* **D10**, 891.
Landshoff, P. V., Polkinghorne, J. C. & Scott, D. M., (1975) *Phys. Rev.* **D12**, 3738.
Landshoff, P. V., Polkinghorne, J. C. & Short, R. D. (1971). *Nucl. Phys.* **B28**, 225.
Landshoff, P. V. and Zakrzewski, W. J., (1969). *Nucl. Phys.* **B12**, 216.
Lee, B. W. & Sawyer, R. F., (1962). *Phys. Rev.* **127**, 2266.
Lévy, M. & Sucher, J., (1969). *Phys. Rev.* **186**, 1656

Mandelstam, S., (1963). *Nuovo Cim.* **30**, 1127, 1148.
Mandelstam, S., (1965). *Phys. Rev.* **137**, B949.
Mason, A. L., (1973). *J. Math. Phys.* **14**, 1601.
Matveev, V., Muradyan, R. & Tavkhelidze, A., (1973). *Nuovo Cim. Lett.* **7**, 719.
McLerran, E. A. & Weis, J. H. (1975). *Nucl. Phys.* **B100**, 329.
Menke, M. N., (1964). *Nuovo Cim.* **34**, 351.
Mueller, A. H., (1970). *Phys. Rev.* **D2**. 2963.

Nash, C. E., (1971). *Nucl. Phys.* **B31**, 419.
Nash, C. E., (1973). *Nucl. Phys.* **B61**, 351.

Olive, D. I. & Polkinghorne, J. C., (1968). *Phys. Rev.* **171**, 1475.

Politzer, H. D., (1974). *Phys. Rep.* **14**, 130.
Polkinghorne, J. C., (1963*a*). *J. Math. Phys.* **4**, 503.
Polkinghorne, J. C., (1963*b*). *J. Math. Phys.* **4**, 1393.
Polkinghorne, J. C., (1963*c*). *J. Math. Phys.* **4**, 1396.
Polkinghorne, J. C., (1963*d*). *Phys. Lett.* **4**, 24.
Polkinghorne, J. C., (1964). *J. Math. Phys.* **5**, 431.
Polkinghorne, J. C., (1965). *Nuovo Cim.* **36**, 857.
Polkinghorne, J. C., (1968). *Nuovo Cim.* **56A**, 755.
Polkinghorne, J. C., (1972). *Nuovo Cim.* **7A**, 555.
Polkinghorne, J. C., (1974). *Phys. Lett.* **B49**, 277.
Polkinghorne, J. C., (1975). *Nucl. Phys.* **B93**, 515.
Polkinghorne, J. C., (1976*a*). *Nucl. Phys.* **B108**, 253.
Polkinghorne, J. C., (1976*b*). *Nucl. Phys.* **B114**, 109.
Polkinghorne, J. C., (1976*c*). *Nucl. Phys.* **B116**, 347.
Polkinghorne, J. C., (1977). *Nucl. Phys.* **B128**, 537.
Preparata, G. (1973). *Phys. Rev.* **D7**, 2973.
Pritchard, D. J. & Stirling, W. J., (1980). *Nucl. Phys.* (in press).

Rothe, H., (1967). *Phys. Rev.* **159**, 1471.

Schwarz, J. H., (1967). *Phys. Rev.* **162**, 1671.
Sivers, D., Brodsky, S. & Blankenbecler, R., (1976). *Phys. Rep.* **23C**, 1.
Stirling, W. J., (1978). *Nucl. Phys.* **B145**, 477.
Sudakov, V. V., (1956). *Soviet Phys. J. exp. theor. Phys.* **3**, 65.
Swift, A. R. & Tucker, R. W., (1970). *Phys. Rev.* D1, 2894.

Ter-Martirosyan, K. A., (1963). *Soviet Phys. J. exp. theor. Phys.* **17**, 233.
't Hooft, G. & Veltman, M., (1972). *Nucl. Phys.* **B44**, 189.
Tiktopoulos, G., (1963). *Phys. Rev.* **131**, 480, 2373.
Tiktopoulos, G. & Treiman, S. B., (1971). *Phys. Rev.* D3, 1037.
Toller, M., (1965). *Nuovo Cim.* **37**, 631.
Tyburski, L., (1976). *Phys. Rev.* D13, 1107.

West, G. B. A., (1970). *Phys. Rev. Lett.* **24**, 1206.

Index

Where a topic is discussed on several consecutive pages reference is given only to the first page.